FEARON'S

Practical Mathematics for Consumers

WORKBOOK

Ronn Yablun

Globe Fearon Educational Publisher
Paramus, New Jersey

Paramount Publishing

Fearon's Pacemaker Curriculum Workbooks

American Government
Basic English
Basic Mathematics
Biology
Careers
Economics
English Composition
General Science
Health
Practical English
Practical Mathematics for Consumers
United States Geography
United States History
World Geography and Cultures
World History

Ronn Yablun holds a B.S. from Northern Illinois University. He is chairman of the math department at Northridge Junior High School in Northridge, California. He is founder of Mathamazement, a learning center for remediation and enrichment in math.

Editor: Sharon Wheeler
Production Editor: Joe C. Shines
Production: Publication Services
Cover Designer: Mark Ong, Side by Side Studios
Cover Photo: Kathleen Campbell/AllStock

ISBN 0-8224-6999-5
Printed in the United States of America

10 9
Cover Printer/NEBC
MA

Table of Contents

What Is Critical Thinking?

Critical thinking—or to put it another way, thinking critically—means putting information to use. Imagine that you are in a supermarket. You want to buy a can of soup, but you are uncertain which brand to choose. As you read the ingredients listed on the label, you review the number of calories in each serving. Perhaps you check the vitamin and fat content. You may also look at the price listed on the shelf. Then you *evaluate* the information and make a choice. You have just used your critical thinking skills. Critical thinking is the act of processing information and using it in a meaningful way.

The activities in this workbook are designed to go along with your Fearon Pacemaker textbook. They ask you to use four different types of critical thinking skills. You have used them all in daily life many times. Now you will be applying these skills to a specific subject.

Some of the activities in this workbook are labeled *Application*. These activities ask you to put information to use by solving a problem or completing a task. For example, imagine that you have just read a section in a United States history book that describes the first battle of the Civil War. You might be asked to *apply* that information by drawing a picture of that battle.

You will find other activities labeled *Analysis*. These activities ask you to look closely at a body of information and to examine all of its parts. You might be asked to *analyze* a map and to answer questions about information the map reveals. You might be presented with an article or an essay and asked to define certain words within it.

Some workbook activities fall under the heading of *Synthesis*. In these activities you will be combining pieces of information to make a whole. You might create a time line by listing events in the order they happened. You might *synthesize* pieces of information into a letter or a diary entry.

You will also find activities labeled *Evaluation*. These exercises ask you to make a judgment about certain information. For example, you might read a statement and decide if it is a fact or an opinion. Remember the supermarket described at the beginning of this page? As a shopper, you were *evaluating* information listed on the soup can label.

Your textbook is a wonderful source of knowledge. By studying it, you will learn a great deal of information in whatever subject the book covers. But the real value of that information will come when you know how to put it to use by thinking critically.

Exercise 1 Analysis — Using an Expense Record

Name ______________________ Date ____________

John is a student at the local junior college. He has a full-time job to pay for his living expenses. Listed below are John's expenses for one month. Use the chart to answer the questions.

	Week 1	Week 2	Week 3	Week 4
Rent	$400.00	—	—	—
Food	$ 50.00	$ 43.00	$ 29.00	$ 48.00
Clothing	—	$ 74.50	$ 29.25	—
Car	—	$195.00	—	—
Gasoline	$ 18.00	$ 12.00	$ 14.00	$ 15.00
Insurance	—	—	—	$ 78.00
Telephone	—	$ 37.55	—	—
Utilities	—	—	$ 42.75	—
Savings	$ 25.00	$ 75.00	$100.00	$100.00
Other	$ 62.50	$ 48.95	$ 38.43	$ 26.25

1. How much money was John able to put in his savings account this month?

   ____

2. What was John's largest expense for the month?

   ____

3. How much did John average per week on gasoline?

   ____

4. Which week did John spend the most on food? Find the difference between the highest and lowest week's amount spent on groceries.

   ____

5. How much money did John need to earn this month to cover his expenses?

   ____

6. Which week was the most expensive for John?

   ____

Exercise 2 Application — Figuring Income

Name ______________________ Date __________

A. Maureen works full-time as a legal secretary. Her roommate Jan also works full-time as a marketing assistant. Maureen's monthly income is $2,250. Jan's weekly income is $575. Answer the following questions.

1. The apartment that Maureen and Jan share costs them $775 per month. They share the cost equally. What is Jan's share of the rent?

   ____

2. Income is based on a four-week month. Who makes more per month, Maureen or Jan? How much more?

   ____

3. Jan's expenses for the month totalled $1,985. After paying these expenses, how much did Jan have left over?

   ____

4. Find Maureen's and Jan's annual income. (Hint: There are 52 weeks in a year.)

   ____

5. Maureen's expenses for one month, not including savings, totaled $1,690. If she plans to save $375 a month, how much will she have left over?

   ____

B. Read and answer each problem.

1. Steven's annual income is $37,900. His monthly expenses are $2,350. Should he be able to save $500 each month?

   ____

2. Tom's monthly income is $2,750. His monthly expenses including savings are $2,200.00. How much does he have left over?

   ____

3. Whose annual income is greater–Steven (problem 1) or Tom (problem 2)? By how much?

   ____

Exercise 3 Application — Comparing Expenses

Name ______________________ Date ____________

A. Write *fixed expense* or *variable expense* for each item.

fixed expense: same cost each month
variable expense: cost changes

1. Telephone bill ____________
2. Rent ____________
3. Gasoline ____________
4. Groceries ____________
5. Car payment ____________
6. Utilities ____________
7. Recreation ____________
8. Savings ____________
9. Clothing ____________
10. Car insurance ____________

B. Use the chart to answer the questions below.

Linda's Monthly Expenses

Rent	$350.00	Gasoline	$ 48.90
Car Payment	$144.00	Clothing	$110.00
Car Insurance	$ 81.00	Recreation	$ 65.00
Utilities	$ 52.25	Savings	$200.00
Credit Cards	$ 60.00	Telephone	$ 54.18
Parking	$ 13.00	Personal Care	$ 22.38
Gifts	$ 33.26		

1. Name the items above that are fixed expenses.

____

2. Name the items above that are variable expenses.

____

3. Did fixed or variable expenses account for the larger part of Linda's budget? How much more?

____

4. Linda's monthly income is $1,650. After covering her expenses, how much does Linda have left?

____

Exercise 4 Application — Planning Monthly Expenses

Name ______________________________ Date ______________

Study the incomes listed in the chart. Next to each income is the amount for fixed expenses. Fill in the amount left for variable expenses. Then answer the questions below.

Job	Monthly Income	Fixed Expenses	Variable Expenses
1. Teacher	$1,400	$ 975	____________
2. Accountant	$2,200	$1,475	____________
3. Waitress	$1,250	$ 850	____________
4. Cashier	$1,335	$1,005	____________
5. Bus Driver	$1,665	$1,227	____________
6. Pilot	$2,000	$1,392	____________

1. Lori works as a secretary for an accounting firm. Her monthly income is $1,475. Her fixed expenses are $945. How much does she have left for variable expenses?

   ____

2. Mark works as a consultant for a computer firm. His weekly income is $425. His monthly expenses are $1,266. How much does he have left for variable expenses? (Hint: There are four pay periods in one month.)

   ____

3. Douglas works as a carpenter. His monthly income is $1,996. His fixed expenses are $1,320. He now pays $775 a month for rent. He wants to move to a new apartment that rents for $900 a month. What will his fixed expenses be if he rents the new apartment? How much will he have left for variable expenses if he rents the new apartment?

   ____

4. Susan is planning on moving to a new city. She has been offered a job at a marketing firm as an assistant for $1,850 per month. Her fixed expenses (not including rent) are currently $445. She has found an apartment she would like to rent for $1,250. She remembers to include variable expenses. Do you think it would be a good idea for Susan to rent this apartment? Why or why not?

   ____

Exercise 5 Application Saving for Variable Expenses

Name ______________________________ Date ______________

Read each problem. Follow the directions carefully.

1. Lisa has $375 left after paying her total expenses for the month. She wants to buy an exercise machine for $660. How much would she have to save for three months to have enough to buy this new exercise machine? For four months? For six months?

   ____

2. Ken has $340 in his savings account. He wants to buy a new stereo system for $800. If he saves $125 a month for the next five months, will he have enough to buy this system without using all of his savings account? If yes, how much will he have left in his savings account after the purchase?

   ____

3. Erica's monthly income is $2,275. Her fixed expenses are $995 and her variable expenses last month were $835. She wants to take a trip to the mountains that costs $425. Does she have enough left over for the trip?

   ____

4. Mr. and Mrs. Swanson both work full-time jobs. Mr. Swanson earns $2,244 per month and Mrs. Swanson earns $1,993 per month. They spend $1,700 on fixed expenses and about $1,500 per month on variable expenses. Which of the following would you recommend they put into a savings account? Circle your choice.

   a. $500 b. $1,000 c. $1,500

5. Wayne's monthly income is $1,975. His fixed expenses, not including rent, are $225, and his variable expenses the past six months have ranged from $500 to $725. He wants to find a new apartment. Which of the following rents per month would you recommend he consider? Circle your choice.

   a. $400 b. $800 c. $1,200

Exercise 6 Analysis — Balancing a Budget

Name ______________________ Date ____________

Study Michael's budget. Then answer the questions below.

Michael's Monthly Budget

Income:		$1,885
Fixed Expenses:	Rent	$ 850
	Utilities	$ 45
	Cable Television	$ 32
	Car Payment	$ 125
	Car Insurance	$ 75
Variable Expenses:	Groceries ($40 per week × 4 weeks)	$ 160
	Clothing	$ 84
	Laundry ($6.75 per week × 4 weeks)	$ 27
	Parking	$ 17
	Gasoline and Auto Expenses	$ 75
	Recreation ($30 per week × 4 weeks)	$ 120
	Savings	$ 275
	Total Expenses:	$1,885

1. How much money will Michael have saved in one year? ______________

2. Find the total fixed expenses. ______________

   Find the total variable expenses. ______________

3. If Michael found a roommate, and split all the apartment expenses (rent, utilities, cable TV), how much money would he add to his variable expenses?

   ____

4. Michael wants to purchase a new color television for $895. How many months will he have to save to have enough to make this purchase? ______________

5. Michael allows $25 each month for auto repairs. How much is he spending on gasoline for his car each week? ______________

Exercise 7 Analysis — Changing a Budget

Name ______________________________ Date ______________

Study the chart. Then answer the questions below.

Budget Changes

Amounts added to budget

- $150 per month added to your savings account for vacation cruise
- $25 per month added to the utilities bill
- $32 per month added to the telephone bill

Amounts subtracted from budget

- $200 per month savings on rent for finding a new roommate
- $16 per month savings on cable television

1. After looking at the changes above, what reason would you give for the increase in the utilities and the telephone bill?

____

2. Which is more, the amounts added to the budget or the amounts subtracted from the budget? How much more?

____

3. On your original budget, you were putting $110 in your savings account each month. You currently have $320 in your savings account. A cruise you want to take will cost $1,750. How many more months will you have to save to be able to afford the cruise?

____

4. You want to take $400 in spending money on the cruise. How many more months will you have to save for this?

____

5. The cruise line is having a 2 for 1 special. If you take a friend, you each travel for $\frac{1}{2}$ fare. How many months will it take you to save for the 2 for 1 fare?

____

Exercise 8 Analysis — Preparing for Emergencies

Name ______________________________ Date ______________

Study the chart. Then answer the questions below.

Monthly Repayment Fund	
Transferred from recreation fund	$25 a month
Transferred from clothing fund	$10 a month
Transferred from savings fund	$35 a month
Transferred from groceries fund	$10 a month
Total	?

1. What is the total monthly repayment fund?

____

2. Why would this become a fixed expense?

____

3. In ten months, this fund will completely repay the debt you owe. How much money do you owe altogether?

____

4. You are going to take $10 from your monthly grocery fund. How much per week will you have to cut back to meet this debt?

____

5. Give two examples of ways you can cut back in your recreation fund to meet your repayment fund.

____

6. You want to go on a weekend ski trip to Utah. The costs are as follows: Round trip transportation—$285, lodging—$85, lift tickets—$72, ski equipment rental—$33, and meals—$100. Using your repayment fund, how many months will it take you to repay the loan if you borrow money for this trip?

____

Exercise 9 Analysis — Budgeting Extra Money

Name ____________________ Date __________

Study the charts. Then answer the questions below.

Changes in Your Income

Income in the old budget:	Additional income from your promotion:	New Income:
$1,885	$375	$2,260

Your Newly Adjusted Monthly Budget

Income:		$2,260
Fixed Expenses:	Rent	$ 850
	Utilities	$ 45
	Cable Television	$ 32
	Car Payment	$ 125
	Car Insurance	$ 75
Variable Expenses:	Groceries ($45 per week × 4 weeks)	$ 180
	Clothing	$ 175
	Parking	$ 17
	Gasoline and Auto Expenses	$ 75
	Recreation ($50 per week × 4 weeks)	$ 200
	Savings	$ 486
	Total Expenses	$2,260

1. Your clothing budget was $84. How much do you now have to spend on clothing?

   ____

2. Your grocery budget was $40 per week. How much more do you now have to spend on groceries per month? ____________________

3. You set aside $275 per month for your savings account. How much per week did this amount increase? ____________________

4. How much of a weekly increase did you receive in income? ____________________

Exercise 10 Application — Adjusting Expenses

Name ______________________ Date ____________

Read the problems. Follow the directions carefully.

1. Maria was spending $750 per month on rent, $55 per month on utilities, and $35 per month on her telephone. If she finds a roommate, her expenses will be cut in half. How much will she save?

   ____

2. Suppose Maria receives a monthly pay increase of $400 (see problem 1). Which is more beneficial—her pay increase or finding a roommate?

   ____

3. Andy just got a promotion that will increase his salary $70 per week. He found a new apartment that rents for $1,150 per month. He now pays $875 per month. Can he afford the new apartment with his raise without affecting the rest of his budget?

   ____

4. Andy (problem 3) decides to take the new apartment. He needs to put down a $575 deposit. He can borrow what he needs, but he wants to pay it back within six months. What is the minimum amount he will need to pay per month in order to repay the debt within six months? (Round your answer to the nearest dollar.)

   ____

5. Jeremy has budgeted $36 per week for recreation. How much has he budgeted altogether for a six-month period? (Hint: There are 52 weeks in 12 months.)

   ____

6. Richard's weekly income is $675. What is his annual income?

   ____

7. What would Richard's monthly income be? (See problem 6.)

   ____

Exercise 11 Analysis — Earning a Salary

Name ______________________________ Date ____________

Fill in the monthly and yearly salary for each job in the chart. Then answer the questions below.

Job	Weekly Salary	Monthly Salary	Yearly Salary
1. Accountant	$725	________	________
2. Mail Carrier	$464	________	________
3. Firefighter	$555	________	________
4. Professor	$512	________	________
5. Dentist	$789	________	________

1. A mail carrier works 40 hours in one week. Find the mail carrier's hourly wage.

   ____

2. A college professor only works four days in a typical work week. How much does a college professor earn each day?

   ____

3. Which person earns the most? Find the monthly and yearly difference between the highest and lowest salaries.

   ____

4. A firefighter puts $350 per month into a savings account. How much does he have left over for expenses?

   ____

5. Whose yearly salary is greater—a firefighter or a professor? How much more?

   ____

6. Whose monthly salary is greater—an accountant or a dentist? How much more?

   ____

Exercise 12 Analysis — Making Overtime

Name ______________________ Date ____________

Fill in the total hours Amy worked each day. Then answer the questions below.

Day	Date	Time In	Time Out	Lunch Time	Total Hours
Mon.	2/1	5 P.M.	11 P.M.	30 minutes	________
Tues.	2/2	9 A.M.	4 P.M.	1 hour	________
Wed.	2/3	1 P.M.	10 P.M.	1 hour	________
Thurs.	2/4	9 A.M.	2 P.M.	30 minutes	________
Fri.	2/5	12 Noon	6 P.M.	30 minutes	________

1. How many total hours did Amy work this week? ______________________

2. Amy is paid $5.50 per hour. How much did she make this week?

   ____

3. Which day did Amy make the most money? How much did she make that day?

   ____

4. Amy is paid overtime if she works more than 40 hours per week. How many more hours will she need to work before she can receive overtime pay?

   ____

5. If Amy works 8 hours Saturday and 6 hours Sunday, will she be eligible for overtime pay?

   ____

6. Amy's overtime pay is one and one-half times her hourly rate. How much will she make in overtime pay if she works the hours in problem 5? (Round your answer to the nearest cent.)

   ____

7. Amy's boss has offered her a raise of $.40 per hour. Find Amy's new wages for her Monday through Friday hours.

   ____

Exercise 13 Analysis — Collecting Tips

Name ______________________________ Date ______________

Fill in the total weekly wages for each profession listed in the chart. Then answer the questions below.

Job	Hourly Wage	Hours Worked	Tips	Weekly Income
1. Waiter	$4.75	40	$225	________
2. Taxi Driver	$6.25	35	$350	________
3. Piano Player	$8.00	15	$ 90	________
4. Bellboy	$4.50	30	$300	________
5. Maid	$5.25	40	$ 75	________

1. Whose weekly income was the greatest? The least? ______________________

2. Does the bellboy make more in wages or in tips? How much more?

   ____

3. Does the waiter make more in wages or in tips? How much more?

   ____

4. Kent drives a taxi during the week and plays piano on the weekend. How much does he make in one week for both jobs? ______________________

5. How much does a waiter make in one year? (Assume the tips are about the same each week.) ______________________

6. Eric works as a waiter and his wife, Ellen works as a maid. What is their combined income for one year? Who makes more? How much more?

   ____

7. A taxi driver makes $29,575 per year including tips. How much is this per month? Round your answer to the nearest dollar. (Hint: There are 12 months in a year.)

   ____

Exercise 14 Analysis — Figuring Expenses

Name ______________________ Date __________

Greg runs a lawn maintenance service. Listed below are his expenses for one month. Use the chart to answer the questions below.

Greg's Expenses for September	
Purchase one new lawn mower	$275.00
Gasoline for lawn tools	$ 38.50
Mobile telephone bill	$195.40
Truck maintenance and upkeep	$144.20
Employee salaries	$850.00

1. What are Greg's total expenses for the month?

   ____

2. Greg's service has 50 clients. Each client pays $45 per month. How much does Greg's service collect each month?

   ____

3. How much profit does Greg's service make this month? (Hint: The profit equals the difference between the expenses and the income.)

   ____

4. Which expense(s) are not regular monthly expenses?

   ____

5. Greg has two part-time employees. If they each make the same salary, how much does each person make?

   ____

6. Look at the information in the chart. Is it possible to determine how much Greg's service spent on gasoline altogether? Why or why not?

   ____

Exercise 15 Application — Working for Wages

Name ________________________ Date ____________

Ellen works as a hotel maid five days a week. She works eight hours a day and her hourly wage is $4.80. Her tips for one month were as follows:

Week 1	**$ 85**
Week 2	**$102**
Week 3	**$ 70**
Week 4	**$ 82**

Ellen is responsible for purchasing her own uniforms at a cost of $33.95. It costs her $2.75 to have each uniform cleaned. Answer the following questions.

1. What were Ellen's total tips for the month?

 ____

2. What is Ellen's weekly wage?

 ____

3. Find Ellen's total income (wages and tips) for the month. (Hint: There are four weeks in one month.)

 ____

4. During the month, Ellen had to buy one new uniform and pay to clean her uniforms six times. How much did Ellen spend on buying and cleaning uniforms for the month?

 ____

5. Ellen puts $15 each week into her savings account. How much does she save in one year? (Hint: There are 52 weeks in a year.)

 ____

6. Ellen is paid one and one-half times her hourly wage for overtime. How much more can she make in one week if she works five hours overtime?

 ____

Exercise 16 Analysis — Figuring Gross Pay

Name ______________________________ Date ______________

Study Derek's time card. Write in the total hours he worked each day. Then use the chart to answer the questions below.

Day	Date	Time In	Lunch	Time Out	Total Hours
Monday	7/8	8:00 A.M.	1 hour	3:00 P.M.	________
Tuesday	7/9	11:00 A.M.	1 hour	7:00 P.M.	________
Wednesday	7/10	8:00 A.M.	1 hour	5:00 P.M.	________
Thursday	7/11	10:00 A.M.	1 hour	7:00 P.M.	________
Friday	7/12	1:00 P.M.	1 hour	10:00 P.M.	________

1. Which day(s) did Derek work the most hours? ______________________

2. How many hours did Derek work altogether for the week?

   ____

3. Derek is paid $6.80 per hour. Find his gross pay for the week.

   ____

4. Derek is entitled to overtime pay if he works more than seven hours in any day. Did he qualify for overtime? ______________________

5. Derek is paid one and one-half times his regular rate of pay for overtime. Find the total he earned just in overtime. ______________________

6. Figure Derek's gross pay for the week including overtime.

   ____

7. If Derek works the same number of hours each week including overtime, how much should he expect to make in one year? (Hint: There are 52 weeks in one year.)

   ____

Exercise 17 Application — Deducting Taxes

Name ______________________________ Date ______________

A. Match Column A to Column B.

Column A	Column B
____ a. Gross pay	1. a reduction in your tax for a person you support
____ b. FICA	2. salary
____ c. Allowance	3. an amount of money withheld
____ d. Net pay	4. Social Security tax
____ e. Deduction	5. take-home pay

B. Heather's paycheck showed that her gross pay for the month was $1,429.33. Answer the questions based on these deductions made from her check:

Federal Income tax	**$268.39**
State Income tax	**$ 79.27**
Social Security tax	**$169.55**
State Disability tax	**$ 6.93**
Credit Union	**$100.00**

1. Which deduction was the greatest? Which deduction was the smallest?

   ____

2. What was the total amount deducted from Heather's paycheck?

   ____

3. How much did Heather have left after these deductions?

   ____

4. If Heather makes the same salary each month, what is her gross salary for one year?

   ____

5. How much federal income tax will be withheld for one year? How much state income tax will be withheld for one year?

   ____

Exercise 18 Analysis — Figuring Net Pay

Name ______________________ Date ____________

Study the income statement. Then answer the questions below.

Name: Tom Reeves	
Monthly Gross Pay:	$1,653.00
Federal Income Tax	$ 172.35
State Income Tax	$ 65.67
FICA	$ 99.29
State Disability	$ 8.17
Health Insurance	$ 85.61
Credit Union	$ 150.00
Net Pay	?

1. Find Tom's total payroll deductions. ____________

2. After all of Tom's deductions, what is his net pay? ____________

3. How much is Tom spending per year on health insurance?

   ____

4. Tom is saving money in a credit union savings account. He began the year with $477. How much should be in this account after one year if he makes no withdrawals?

   ____

5. What is Tom's total gross and net pay for one year? ____________

6. How much does Tom have deducted annually for federal income tax?

   ____

   How much does Tom have deducted annually for state income tax?

   ____

7. Tom has decided to reduce his credit union deduction by $25 per month. How much will this add to Tom's net pay annually?

   ____

Exercise 19 Analysis — Figuring Income

Name ______________________ Date __________

Study the earnings statement. Then answer the questions below.

Action Marketing Corp.
Pay Period Ending: 12/29

Employee: Frank Padrillo
Social Security Number: 999-00-9090

Gross Pay	Federal Tax	State Tax	FICA	Health Ins.
$975.00	$115.35	$17.48	$83.52	$111.00

Disability Insurance	Union Dues	Credit Union	Net Pay
$8.66	$27.25	$75.00	?

1. Frank gets paid twice each month. His credit union deduction goes directly into a savings account. How much does he put into this account in six months? In one year?

   ____

2. Find Frank's net pay.

   ____

3. What is Frank's net pay for six months? For one year? (Hint: Remember Frank gets paid twice each month.)

   ____

4. How much federal tax does Frank have withheld in one year?

   ____

5. How much state tax does Frank have withheld in one year?

   ____

6. Frank is thinking about reducing his credit union deduction from $75 to $50. How much would this increase his net pay over six months? In one year?

   ____

7. What percent of Frank's gross pay is his net pay?

   ____

Exercise 20 Application — Starting a New Job

Name ______________________________ Date ______________

Darren is beginning a new job as an assistant manager in a men's clothing store. The starting salary is $5.50 an hour. The hours are from 9:00 A.M. to 6:00 P.M. Monday through Friday. There is a one-hour unpaid lunch break each day. Answer the following questions.

1. What is Darren's gross pay for one week? For one year? (Hint: There are 52 weeks in one year.)

   ____

2. Darren's deductions are 28% of his gross pay. How much is deducted weekly from his paycheck?

   ____

3. What is Darren's net pay for one week? For one year?

   ____

4. What percent of his gross pay is his net pay? (Hint: See problem 1.)

   ____

5. Darren has been offered to participate in the company's health plan at a cost to him of $12.75 per week. If Darren decides to participate, what will his weekly net pay be?

   ____

6. Darren has decided to join the company health plan. What percent of his gross weekly salary are his deductions? (Round your answer to the nearest whole percent.)

   ____

7. Darren's weekly fixed and variable expenses total $122 per week. How much does Darren have left each week after covering his expenses?

   ____

Exercise 21 Analysis — Using a Bank

Name ________________________ Date ____________

Compare the three banks in the chart. Then answer the questions below.

Type of Account	Commercial Bank	Savings Bank	Credit Union
Savings	6.5% interest	6.75% interest	6.8% interest
Checking	$9.00 per month or free with $2,000 minimum	$7.50 per month or free with $2,500 minimum	$8.00 per month or free with $1,000 minimum
Loans	9.75% interest	8.75% interest	7.5% interest

1. Which bank pays the highest rate of interest on a savings account?

____

2. If Eric needed to borrow $2,500 for a down payment on a car, which bank should he go to? Why?

____

3. Eric wants to keep no more than $500 in his checking account. With which bank should he open a checking account? Why?

____

4. Suppose Eric keeps a balance of $1,500 in his checking account. With which bank should he open a checking account? Why?

____

5. Eric's monthly take-home pay is about $1,500. His fixed and variable expenses are $900 per month altogether. He saves $100 each month. He puts the rest of his paycheck into a checking account. Eric is considering changing banks. He wants to keep his checking and savings account at the same bank. Which bank should he choose? Why?

____

____

Exercise 22 Analysis — Banking Services

Name ______________________ Date __________

Compare the three banks in the chart. Then answer the questions below.

Account Services	Commercial Bank	Savings Bank	Credit Union
Checking	$ .08 per check $9.00 per month	$ .07 per check $7.50 per month	$ .07 per check $8.00 per month
Savings	6.5% interest	6.75% interest	6.8% interest
ATM	$.50 per transaction	$.40 per transaction	not available
Money orders	$2.00 each	$2.25 each	$1.75 each
Safety deposit box	$6.00 per year	$3.50 per 6 months	not available

1. Kelly wants to open a checking and savings account at a local bank. She expects to use the automated teller three or four times a month. Which bank would offer her the most reasonable fees for this service?

____

2. Kelly plans to write about twenty checks per month. What is the total amount she will pay for this service at the Commercial Bank? At the Savings Bank?

____

3. Which bank offers the best rate for a safety deposit box?

____

4. Kelly has $300 to open a savings account. How much interest will she earn in one year on a savings account at the Commercial Bank? At the Savings Bank? At the Credit Union?

____

Exercise 23 Application — Figuring Interest

Name ______________________ Date __________

A. Find the interest earned on each of these accounts.

1. What is 7% yearly interest of $1,500? What is the monthly interest? (Hint: there are 12 months in one year.)

____

2. What is 11% yearly interest of $2,000? What is 25% (one-quarter) of the yearly interest?

____

3. What is 5.5% yearly interest of $800? What is 50% (one-half) of the yearly interest?

____

4. What is 7.75% yearly interest of $1,000? What is 75% (three-quarters) of the yearly interest?

____

5. What is 10.5% yearly interest of $750? How much interest is earned after six months?

____

B. Read each question. Follow the directions carefully.

1. You put $652.48 into a savings account that pays 8% interest. How much interest will you earn in

   one year? ____________ two months? ____________

   three months? ____________

2. You put $1,245.25 into a savings account that pays 7% interest. How much interest will you earn in

   one month? ____________ one year? ____________

3. You put $5,200 into a savings account that pays 9.75% interest. How much interest will you earn in

   one year? ____________ four months? ____________

Exercise 24 Analysis Comparing Savings Accounts

Name ______________________________ Date ______________

Compare the three banks in the chart. Then answer the questions below.

	Commercial Bank	Savings Bank	Credit Union
Passbook account*	6.5% interest	6.75% interest	6.8% interest
CD Account*	3 months—7% interest 6 months—8% interest 1 year—9% interest	3 months—7.5% interest 6 months—8.5% interest 1 year—9% interest	3 months—7.5% interest 6 months—8.25% interest 1 year—9.25% interest

*All interest is based on 12 months.

1. Victor has $1,500 he wants to put in a savings account. If he can only leave it in for six months, which bank should he use? What kind of an account should he place the money in? How much interest will he earn on this account?

____

2. Victor has decided to split the money so he can put some of it in a long-term CD. He is going to place $1,000 in a one-year CD and the rest in a three-month CD. Which bank will pay him the most interest for this type of investment? How much interest will he earn altogether?

____

3. Victor is planning to put $250 in a passbook account. Which bank will pay the highest interest rate? How much interest will he earn in one month? In six months? In one year? (Round your answers to the nearest cent.)

____

____

4. Victor decided to put the entire $1,500 into a six-month CD. Which bank will pay him the best rate over this period of time? How much interest will he earn on this account?

____

5. If Victor were to put $500 into a passbook savings account for one year and $1,000 into a one-year CD, which bank would pay him the best rate? How much interest would he earn altogether?

____

Exercise 25 Application — Investing Money

Name ______________________________ Date ______________

Brad works for Miller Plumbing. His monthly paycheck is $1,957.48. His monthly fixed and variable expenses total $1,326.25. His bank offers a passbook savings account that pays 7.5% interest. They offer a six-month CD that pays 8% interest, and a one-year CD that pays 8.25% interest. They also charge $8 per year for a safety deposit box and $1.50 for each money order. Answer the following questions.

1. How much income does Brad have after he has paid his monthly expenses?

   ____

2. Brad decided to rent a safety deposit box. During that same month, he also purchased 12 money orders. What is the total amount he spent on fees for a safety deposit box and the money orders?

   ____

3. Brad has decided to invest $250 in a one-year CD. How much interest will he earn in one year?

   ____

4. Brad has decided to put $550 in a passbook savings account. How much will be in the account at the end of one year?

   ____

5. Brad's bank charges $8.50 per month for a checking account with a $.10 per check charge. Brad writes 15 checks per month. What is the total charge for Brad's checking service?

   ____

6. Circle the account in which Brad would earn the most interest.

   a. $400 in a passbook savings account for one year

   b. $500 in a six-month CD

Exercise 26 Analysis — Making Deposits

Name ______________________________ Date ______________

Study the deposit slip. Then answer the questions below.

CHECKING ACCOUNT DEPOSIT SLIP

DATE January 2 19 93

Michael Stewart

SIGN HERE FOR CASH RECEIVED

CASH	Currency	$125.	00
	Coin	$25.	00
List Checks Singly			
	16-66	$167.	60
TOTAL		$317.	60
LESS CASH RECEIVED		$100.	00
NET DEPOSIT		$217.	60

16-66
1220

USE OTHER SIDE FOR ADDITIONAL LISTING

BE SURE EACH ITEM IS PROPERLY ENDORSED

National Bank of Chicago
North Park Drive
Chicago, Illinois

1: 1220 00851 1:7201...20355...0390911

1. How much currency and coin is Michael depositing altogether? ______________
2. What is the amount of the check that Michael is depositing? ______________
3. What is the date of this deposit? ______________
4. Why did Michael sign his name under the date? ______________
5. How much cash is Michael receiving from the deposit? ______________
6. What is the bank number on the check? ______________
7. How much is Michael depositing into his checking account? ______________
8. When should Michael sign the deposit slip for cash? Why?

____

____

Exercise 27 Analysis — Writing Checks

Name ______________________ Date __________

Study the check. Then answer the questions below.

Mark Johnson
12457 Gerry Place
Los Angeles, Ca 90002

202

② ______________ 19____

16-66/1220

PAY TO THE ORDER OF ______①______ $ ___③___

______④______ DOLLARS

National Bank
9888 Wilshire Blvd.
Los Angeles, CA 90002

Memo ______________ ______⑤______

1:1220 008511:7201...20355...0395711

1. On which numbered line should Mark sign the check? ______________
2. What is the memo line for? ______________
3. On which numbered line should Mark put the date? ______________
4. What should Mark write on line 4? ______________
5. What do the numbers under the memo line mean? ______________
6. What is the name of the bank where Mark has this checking account?

   ____
7. What should Mark write on line 1? ______________
8. On this check, what do the numbers 16-66/1220 mean? ______________
9. What is the check number? ______________

Exercise 28 Analysis Balancing a Checking Account

Name ______________________ Date __________

Study the check register. Then answer the questions below.

CHECK NO.	DATE	CHECKS ISSUED TO OR DESCRIPTION OF DEPOSIT	AMOUNT OF CHECK (–)		T	AMOUNT OF DEPOSIT (+)		BALANCE	
								632	37
478	8/23	SouthCo Electric	47	83				A	
	8/23					375	00	959	54
479	8/23	John's Market	63	37				896	17
		Groceries							
480	8/25	Bell Telephone	B					843	95
		(July bill)							
C	8/28	First National Bank Loan	45	00				798	95
	8/28					D		898	95

1. What period of time does this check register cover?

____

2. Find the "balance" column. Find letter A. What balance amount belongs here?

____

3. Find the "amount of check" column. Find letter B. What was the amount of the check written to Pacific Bell Telephone?

____

4. Find the "check number" column. Find letter C. What is the check number for the check written to the First National Bank?

____

5. Find the "amount of deposit" column. Find letter D. What was the the amount of the deposit made?

____

6. From the check register, can you tell on what date the last deposit was made?

____

Exercise 29 Analysis — Reading Bank Statements

Name ______________________ Date ____________

Study the bank statement. Then answer the questions below.

National Bank of Chicago
North Park Drive Chicago, Illinois

Michael Green
4267 Milwaukee Avenue
Chicago, Illinois 60023

Closing Date: 11/29/94

Beginning Balance: $523.77

Checking Account Number 21557643

CHECKS			DEPOSITS	
Check Number	**Date Paid**	**Amount**	**Date**	**Amount**
331	11/14/94	43.35	11/08/94	$150.00
332	11/17/94	56.02	11/22/94	$873.26
330	11/20/94	133.55		
333	11/22/94	22.27		
335	11/27/94	77.91		
334	11/28/94	14.44		

OTHER CHARGES
Service Charge 11/29/94 8.90

Ending Balance ?

1. Find the ending balance. ____________
2. What is Michael Green's checking account number? ____________
3. How many checks were written? Find the total amount of these checks.

____

4. How many deposits were made? Find the total of these deposits.

____

5. If the service charge were listed with the checks and deposits, which column would it be listed under? Why?

____

6. Find the difference between the beginning and ending balance. ____________

Exercise 30 Application Balancing Accounts

Name ______________________________ Date ____________

Read each question. Follow the directions carefully.

1. Sara has a current balance of $237.53 in her checking account. She wrote checks for $56.09, $25.66, $112.67, and $58.38. What error has Sara made?

   ____

2. Beth made three deposits for $234.12, $44.69, and $68.25. She wrote checks for $59.75, $77.27, $199.26, and $21.23. If she started the month with $112.53, what will be the balance in her account after these transactions?

   ____

3. Lauren's checking account had a balance at the beginning of the month of $344.26. She deposited her paycheck in the amount of $1,326.55. She then wrote a check for her rent in the amount of $850. What is the balance in her account now?

   ____

4. Nancy just received her bank statement on her checking account. At the bottom of the statement, she notices that the ending balance is −$23.12. What does the minus sign in front of the amount tell her about her account?

   ____

5. Dennis wrote a check to pay for a purchase at the drug store and forgot to enter the amount in his check register. Circle the sentence that tells what Dennis should do first.

   a. He should wait until his statement comes at the end of the month.

   b. He should try to find the register receipt from the drug store.

   c. He should estimate the amount of the check.

   d. He should keep a higher balance in his account until he can find the amount.

6. Terry's bank will not charge him a service charge for his checking account as long as he maintains a balance of $1,000. The bank charges $9 per month and 11¢ per check. If Terry can maintain the $1,000 balance, how much will he save if he writes an average of 23 checks each month?

   ____

Exercise 31 Application Finding an Apartment

Name ______________________________ Date ______________

Maria works for a messenger service in the downtown area. Her company provides her with a car to make deliveries, but the car must be returned at the end of the day. She gets paid $6.80 per hour. Answer the following questions.

1. Maria has found an apartment she likes that is 18 miles from the office. The bus fare will cost her $1.25 each way. She works five days each week. She can buy a bus pass for $45 that will last four weeks. Which would be the least expensive for her? By how much?

   ____

2. Maria has a friend who works for the same company. She can carpool if she pays $10 per week toward the cost of gasoline. Would she be better off riding the bus or participating in the carpool?

   ____

3. Maria has found a used car she can buy for $2,400. Her monthly payment would be $74.85 for 36 months. She figures the gas will cost her $14 per week. What other costs will Maria need to consider before she makes this purchase?

   ____

4. Maria has figured she should be able to spend $450 per month on rent. Her monthly income is about $1,250 including overtime pay. What percent of her income has she allowed for rent?

   ____

5. Maria has found another apartment in the same area that she likes. The first apartment she found rents for $425 per month. The second apartment rents for $435 per month. The first apartment has one bedroom and one bathroom. The second apartment has two bedrooms and one bathroom. Which apartment is the best deal?

   ____

Exercise 32 Analysis Reading Ads

Name ______________________ Date __________

Jody is looking for an apartment to rent. Answer the questions based on this ad she found in the newspaper.

For rent: Lge. 2 BR, 2 BA apt. AEK,FP, pool. $450 mo. incl. all util. nr. shopping, schools Avail. 4/1; 1st, sec. & cleaning req'd. 555-2346 or 555-2347 mess.

1. How many bedrooms and bathrooms does this apartment offer?

____

2. How much will Jody pay for utilities?

____

3. When can Jody move into this apartment?

____

4. What would Jody have to pay in advance if she decides to rent this apartment?

____

5. Jody has called 555-2346, and there is no answer. What should she do?

____

6. What "special features" does this apartment offer?

____

7. What does Jody know about the apartment's location?

____

8. What does Jody know about the size of the apartment?

____

Exercise 33 Application — Paying a Deposit

Name ______________________ Date __________

Robert moved into a new apartment last month. The rent is $475 per month. He paid first and last month's rent, a $400 security deposit, and a $175 cleaning fee. Answer the following questions.

1. How much in fees and deposits did Robert have to pay altogether?

   ____

2. When Robert decides to move, he can do the final cleaning. How much will he receive for this?

   ____

3. Robert lost two keys. He is charged $25 for each key. How much will he be charged for both keys? This amount will be deducted from which fee that Robert paid? How much of this fee should Robert expect to receive when he moves?

   ____

4. When Robert decides to move, how much will he have to pay for the final month's rent? How do you know this?

   ____

5. Robert's utilities are $40 a month, and he has a cable TV subscription for $15.75 a month. What are Robert's total payments for the apartment?

   ____

6. After living in the apartment for six months, Robert receives a notice that the rent will be increased by $35 per month. How much rent will he now have to pay?

   ____

Exercise 34 Application — Signing Leases

Name ______________________________ Date ______________

A. John and Mark are renting a two-bedroom apartment together. The electric bill came to $48.80, and the gas bill came to $12.22. The telephone bill was $48.28. They signed a one-year lease for the apartment starting November 1, 1994 for $860 per month for one full year. Answer the following questions.

1. How much is John's share of the rent?

   ____

2. How much is John's share of the utilities including the telephone bill?

   ____

3. What is the total monthly cost for Mark including rent and utilities?

   ____

4. On February fifth, Mark and John received a notice that the rent was going to be raised $25 per month. Is this legal?

   ____

5. Their lease states that an additional roommate would add $100 to the monthly rent. If they decide to take on another roommate and split the rent evenly, how much will each person pay?

   ____

B. List four items that would be covered in a lease.

____

____

____

____

Exercise 35 Application — Paying Rent

Name ______________________________ Date ______________

Lisa and Julie have just signed a two-year lease on a two-bedroom apartment. The rent will be $775 the first year and $825 the second year. Lisa works at a fabric house as a designer for $9.40 an hour, and Julie works at a dentist's office as an office manager for $8.10 an hour. They expect their monthly telephone bill to be about $50 and their utilities to cost about $60 per month. Answer the following questions.

1. What is Julie's half of the rent for the first year? For the second year?

   ____

2. What is Lisa's share of the monthly utilities and telephone bill?

   ____

3. Julie works 40 hours each week. Find her weekly gross income. How much does she make in a four-week (one-month) period?

   ____

4. Julie doesn't want to spend more than 30% (.30) of her gross salary on rent. What is 30% of her gross salary? Is she within her limit on this new apartment?

   ____

5. Lisa works 35 hours each week. Find her weekly gross income. How much does she make in a four-week (one-month) period?

   ____

6. Lisa doesn't want to spend more than 20% (.20) of her monthly gross salary on rent. What is 20% of her monthly gross salary? Is she within her limit on this new apartment?

   ____

7. Lisa's employer deducts 30% of her monthly gross salary for taxes and other deductions. After paying for her share of the apartment expenses and rent, how much will she have left for other living expenses?

   ____

Exercise 36 Application — Buying Furniture

Name ______________________________ Date ______________

Joanne is moving into a one-bedroom apartment. She needs to budget her salary so she can furnish the apartment without going into debt. She makes $9.25 an hour as a medical assistant. Answer the following questions.

1. Joanne works 40 hours each week. Find Joanne's monthly (four-week) gross income.

   ____

2. Joanne's employer deducts 30% (.30) of her gross salary for taxes and other deductions. How much is deducted monthly from her pay? How much pay does she receive monthly?

   ____

3. Joanne knows that her other expenses (fixed and variable) will cost her $825 including the apartment. How much will she have left each month to spend on furnishings?

   ____

4. Joanne went to a garage sale to buy a kitchen dining set. She found a table for $35 and four chairs for $13 each. How much will the complete set cost?

   ____

5. Joanne went to a "finish it yourself" furniture store and found a rocking chair for $46. It also cost her $8.75 for supplies to finish the chair. How much will the rocking chair cost her altogether?

   ____

6. Including what she has spent on the kitchen set and the rocking chair, how much has she spent on furnishings so far?

   ____

7. How much does Joanne have left this month after buying the kitchen set and the rocking chair? (See problem 3.)

   ____

Exercise 37 Application — Using Layaway

Name ______________________________ Date ____________

Larry needs to buy furnishings for his new apartment. He is a construction worker and makes $12.24 per hour. He works a forty-hour week and sometimes has the opportunity to put in overtime hours at time and a half. Answer the following questions.

1. How much does Larry make in a four-week period (gross salary)?

   ____

2. After 30% (.30) deductions, how much of his salary does he have left?

   ____

3. Larry's fixed and variable expenses total $975 per month. How much does he have left after these expenses?

   ____

4. Larry has set a limit of 50% of the income he has left after expenses to spend on new furnishings. How much will he have this month to spend?

   ____

5. Larry found a new dining room set that costs $420. He can put it on layaway if he makes a down payment of 25% and pays it off in six equal monthly payments. How much will his down payment be? How much will his monthly payments be?

   ____

6. Larry found a new loveseat on sale for $350. He can make a down payment of $75. What is the balance left on the loveseat?

   ____

7. Larry has decided to make five monthly payments on the balance of the loveseat. How much will each payment be? What will Larry's total monthly payments be for the dining room and the loveseat?

   ____

Exercise 38 Application — Buying on Credit

Name ______________________ Date __________

Tim is 28 years old and has worked as a bus driver for the Metropolitan Bus System for the past five years. He makes $8.80 per hour and works a 40-hour week. He pays $675 per month rent. He is single and has lived in his present apartment for two years. His total monthly fixed and variable expenses total $925. Circle the correct answer below each question.

1. Tim is filling out a credit application. One question on the application asks his annual salary. Which box should he check?

   a. Under $15,000   b. $15,000–$25,000   c. $25,000–$35,000

2. The credit application states: Previous employer (if under three years at present job). Should Tim fill in this section of the application?

   a. yes   b. no   c. Tim's employer should fill it in.

3. Which of the following would be the best credit reference for Tim to write on his credit application?

   a. his employer   b. his brother   c. the telephone company

4. Which of the following should Tim write on his credit application to verify his employment?

   a. Southwestern Bell   b. Metropolitan Bus System   c. Credit Union

5. Which sections of a credit application will Tim need to fill out?

   a. Previous address if you have lived at your present address for less than three years.

   b. If married, fill in this information about your spouse.

   c. How long have you been employed at your present job?

   d. If under the age of 21, have your parent(s) or legal guardian fill out this section.

Exercise 39 Application Buying Household Items

Name ______________________ Date __________

A. Tracy is moving into a new apartment. She needs to purchase household necessities. She has set a limit for herself of $75 per week to make these purchases. Answer the following questions.

1. For the kitchen, Tracy made a list that includes items on sale at these prices:

   cleanser—$.69
   dishwasher soap—$2.09
   sponges—$1.99
   kitchen towels—$1.99
   trash bags—$1.79

   She has set $30 as her limit to make these purchases. What will the total cost be if she buys three towels and one of each of the other items? Is she within her limit?

   ____

2. For the bathroom, Tracy made a list that includes items on sale at these prices:

   bath towels—$4.99
   hand towels—$2.99
   hand soap—$1.49
   toothpaste—$1.19
   hairbrush—$2.59
   box of tissue—$.99

   She wants to buy three bath towels, three hand towels, one bottle of hand soap, one tube of toothpaste, one toothbrush, one hairbrush, and two boxes of tissue. What will the total cost of these items be?

   ____

B. Imagine you are moving into a new apartment. List five household products for the kitchen below. Use the supermarket ads from a newspaper to compare prices of the five items. Write the prices next to each product.

____

____

____

____

____

Exercise 40 Application — Furnishing a House

Name ______________________________ Date ______________

Zack is moving into a two-bedroom house he has rented for $800 per month. He makes $600 per week (gross) as a computer consultant. He needs to furnish the house and also purchase necessities. He already has his bedroom furniture and a sofa for the living room. Answer the following questions.

1. Zack's employer deducts 30% from his monthly paycheck. After paying his rent, how much does Zack have left for other expenses?

   ____

2. Zack's other fixed and variable expenses, besides rent, total $500. How much does Zack have left after these monthly expenses are paid?

   ____

3. Zack found a dining room set for $650. He can put it on layaway if he makes a 25% down payment. He must also pay it off within six months. How much will he have to pay as a down payment? How much will each of the six monthly payments be?

   ____

4. Zack went shopping for some basic kitchen necessities. He purchased a mop for $3.99, a broom for $2.79, dish soap for $2.19, sponges for $1.19, paper towels for $.79 and a dish drainer for $5.99. How much did he spend altogether?

   ____

5. Zack went to a flea market and purchased these items: a table lamp for $12, an end table for $15, a swivel rocker for $55, a TV cart for $15 and a pair of stools for $26. How much did he spend altogether?

   ____

6. So far, Zack has purchased basic kitchen necessities, purchased furnishings at the flea market, and made a down payment on a dining room set. How much has he spent this month?

   ____

Exercise 41 Analysis — Comparing Food Prices

Name ______________________ Date __________

Brett and Sally went shopping at the supermarket. They took this local newspaper ad to help them decide what to buy.

Smith's Market
Super Low Prices
This Week's Specials

Sweet corn	3 cans only \$1.89
Wax Beans	4 cans only \$1.37
Asparagus	2 cans only \$0.89
Carrots	5 cans only \$1.75

1. Sally wants to buy one can of sweet corn. How much will she pay for one can?

   ____

2. Brett wants to purchase two cans of wax beans. How much will he pay for the two cans? ______________________

3. Brett and Sally decided to buy three cans of carrots. How much will they pay for the three cans? ______________________

4. Brett wants to find out which can of vegetables is the best buy. Which operation should he use: addition, subtraction, multiplication, or division?

   ____

5. Sally decided to use her calculator to help Brett find the price per can of each vegetable. To find the cost per can of asparagus, Sally should:

   a. Enter __________ into her calculator.

   b. Press the __________ key.

   c. Enter __________ into the calculator.

   d. Press the __________ key.

   e. Her answer should be __________.

6. Which of the four canned vegetables is the best buy per can? ______________

Exercise 42 Application — Measuring Ingredients

Name ______________________________ Date ______________

You are planning to shop for the ingredients to make soup for eight people. Read the recipe. How much of each item would you need? Suppose you wanted to make enough for only two people. How much of each item would you need? Fill in the columns below.

Vegetable Soup

$1\frac{1}{2}$ pounds of fresh tomatoes, chopped
2 medium-sized potatoes, cut into small pieces
$\frac{1}{2}$ cup of green beans, cut into 1-inch pieces
2 carrots, cut into small pieces
3 stalks of celery, cut into small pieces
1 large onion, chopped
1 teaspoon thyme
4 cups of water

Place all of the ingredients in a large cooking pot. Bring to a boil. Then simmer for at least one hour. Serves four people.

Item	Eight Servings	Two Servings
tomatoes	________	________
potatoes	________	________
green beans	________	________
carrots	________	________
celery	________	________
onion	________	________
thyme	________	________

Exercise 43 Analysis — Reading Labels

Name ______________________ Date __________

Study the food label. Answer the following questions.

Ingredients

enriched rice, vegetables, cheddar, romano, and parmesan cheeses, partially hydrogenated vegetable oil, salt, whey, buttermilk, nonfat milk, modified food starch, chicken fat, natural flavor, chicken broth, silicon dioxide, sodium phosphate, annatto, turmeric, sugar, chicken meat, soy flour, sulfites.

Nutrition Information Per Serving

Serving size	32g	Servings per package	4	Calories	120
Protein	3g	Carbohydrate	21g	Sodium	380mg

Percentage Of U.S. Recommended Daily Allowance (U.S. RDA):

Protein	6	Vitamin A	*	Vitamin C	*	Thiamine	8
Riboflavin	6	Niacin	6	Calcium	6	Iron	4

*contains less than 2 percent of the U.S. RDA of these nutrients.

1. Look at the nutrition information on this label. What is the serving size? How many servings are there in this package?

   ____

2. How many milk products are contained in this package? Name them.

   ____

3. Look at the percentage of U.S. RDA information on this label. What does the * after Vitamin A mean?

   ____

4. How many grams of carbohydrate are in the package?

   ____

5. Look at the percentage of U.S. recommended daily allowance information on this label. What does the 8 after Thiamine mean?

   ____

6. On this label, the abbreviation "mg" is used for sodium. What does it mean?

   ____

Exercise 44 Analysis — Comparing Products

Name ______________________ Date ____________

Fernando compared canned, frozen, and fresh food prices at his local market. Compare the prices below. Then answer the following questions.

Item	Canned	Frozen	Fresh
Cherries	$2.09 per 12 ounces	$1.29 per 8 ounces	$2.19 per pound
Broccoli	$1.19 per pound	$ .89 per 10 ounces	$1.29 per pound
Carrots	$ .49 per 8 ounces	$ .69 per 12 ounces	$ .79 per pound
Blueberries	$2.39 per pound	$2.09 per 12 ounces	$1.69 per 8 ounces

1. There are 16 ounces in 1 pound. How much will Fernando pay for one pound of canned carrots?

   ____

2. Fernando wants to buy 2 pounds of cherries. Will his best buy be canned, frozen, or fresh? How much will he pay altogether for the 2 pounds?

   ____

3. Fernando has broccoli on his shopping list. What is the cost per pound of frozen broccoli? He wants to buy $\frac{1}{2}$ pound altogether. Will his best buy be canned, frozen, or fresh? How much will he pay altogether for the $\frac{1}{2}$ pound of broccoli?

   ____

4. Fernando has decided to buy fresh blueberries. How much will he pay for 1 pound of fresh blueberries?

   ____

5. Fernando prefers to buy fresh vegetables. He wants to buy $\frac{1}{2}$ pound of carrots and $\frac{3}{4}$ pound of broccoli. How much will this cost him altogether?

   ____

6. Fernando prefers to buy frozen fruits. He wants to buy 1 pound of cherries and $\frac{1}{2}$ pound of blueberries. How much will this cost him altogether?

   ____

Exercise 45 Analysis — Shopping for the Best Buy

Name ______________________________ Date ______________

Theresa is shopping for the lowest-priced items at the supermarket. Answer the following questions.

1. Red apples are on sale at 2 pounds for $1.89. Green apples are on sale at 3 pounds for $2.79. Which would be the better buy? Theresa wants to buy 4 pounds. How much will it cost her?

   ____

2. Frozen lemonade is sold for $1.19 per 12-ounce can. One can makes 2 quarts of lemonade. Fresh lemonade is on sale for $1.09 a half-gallon. There are 4 quarts in a gallon. Which should Theresa buy?

   ____

3. Chicken thighs are on sale for $.99 per pound, and chicken wings are on sale for $.79 per pound. Theresa has decided to purchase 3 pounds of chicken thighs and 2 pounds of chicken wings. How much will this cost her altogether?

   ____

4. Strawberry ice cream is on sale this week. Brand A comes in an 8-ounce carton for $1.29. Brand B comes in a 12-ounce carton and is on sale for $1.79. Which brand should Theresa buy?

   ____

5. Milk comes in $\frac{1}{2}$ gallon and 1 gallon cartons. The 1 gallon carton costs $2.49, and the $\frac{1}{2}$ gallon carton costs $1.39. Which item is the better buy?

   ____

6. Bananas are on sale for $.39 per pound, and strawberries are on sale for $.59 per pound. Theresa wants to buy 3 pounds of bananas and 2 pounds of strawberries. How much will these two items cost altogether?

   ____

Exercise 46 Analysis — Studying Food Groups

Name ________________________________ Date ______________

Lena tries to provide her family with nutritious meals. She and her husband have three children, ages 9, 13, and 16. Answer the following questions.

1. Lena knows that each member of her family needs 5 to 7 ounces of meat a day. Lena is shopping for seven days' meals. What is the minimum quantity of meat Lena needs to purchase for her family to meet this need?

   ____

2. Each member of Lena's family should have 6 to 11 servings from the bread and cereal group each day. On Tuesday, she is planning to make spaghetti with meatballs. This will meet the requirements from two food groups. Name them.

   ____

   ____

3. Lena wants to be sure that her children drink enough milk. Nine- to twelve-year-olds need at least 3 cups daily and teenagers need 4 or more cups daily. What is the minimum requirement her children need daily? What is the weekly total?

   ____

4. According to the USDA, each member of Lena's family should have 2 to 4 servings from the fruit group daily. How many servings will her family need altogether for an entire week to meet the minimum requirement set by the USDA?

   ____

5. The USDA recommends that each person have 3 to 5 servings of a dark green or deep yellow vegetable at least every other day. How many servings of these types of vegetables does Lena need to buy to meet the minimum requirement for her entire family for two weeks?

   ____

Exercise 47 Analysis Counting Calories

Name ______________________ Date __________

Study the list of foods and the number of calories each contains. Answer the questions below.

Calorie Count

Food	Calories	Food	Calories	Food	Calories
Apple	81	Ground Beef (3 oz.)	204	Hamburger Bun	114
Carrot	31	Cheese (1 oz.)	114	Chicken (4 oz.)	278
Diet Cola	6	Sugar Cookie	98	Cucumber (1 oz.)	4
Egg	92	Tuna (3 oz.)	169	Hot Dog	183
Grapefruit ($\frac{1}{2}$)	37	Ham (3 oz.)	106	Ice Cream (1 cup)	377
Lettuce ($\frac{1}{2}$ head)	11	Milk (1 cup)	121	Peanut Butter (1 T)	52
Orange	62	Pancakes (1)	60	Peach	37
Apple Pie (slice)	283	Popcorn (1 cup)	25	Potato Chips (10)	105
Baked Potato	145	Rice ($\frac{1}{2}$ cup)	132	Shrimp (3 oz.)	84
Spinach ($\frac{1}{2}$ cup)	29	Strawberries ($\frac{1}{2}$ cup)	23	Turkey (3 slices)	145
Tomatoes ($\frac{1}{2}$ cup)	24	Veal (3 oz.)	179	Yogurt ($\frac{1}{2}$ cup)	115

1. Before leaving for school in the morning, Jeff decided to make himself breakfast. He made 3 pancakes, 2 eggs, and $\frac{1}{2}$ a grapefruit. He also had a glass of milk (2 cups). How many calories did Jeff have altogether for breakfast?

   ____

2. Jeff is going to have a hamburger, potato chips, and milk for lunch. The hamburger is 9 ounces of ground beef and comes with a bun. The package of potato chips has 50 chips, and the glass of milk is equal to 2 cups. Find the total number of calories in Jeff's lunch.

   ____

3. After eating lunch, Jeff was still hungry so he decided to have a slice of apple pie and another glass of milk (2 cups). How many calories did Jeff's dessert total?

   ____

4. For dinner, Jeff had chicken (12 ounces), a baked potato, and a diet cola. How many calories did Jeff have for dinner?

   ____

5. What is the total number of calories Jeff consumed for the day?

   ____

Exercise 48 Application Making Choices

Name ______________________ Date __________

Patrick enjoys working out and keeping in shape. He is also very particular about what he eats. Every week he reads through the food section of the newspaper to put together his grocery list. Answer the following questions.

1. One ounce of American cheese contains 106 calories and 8.9 grams of fat. One ounce of cheddar cheese contains 114 calories and 9.4 grams of fat. American cheese comes in 16-ounce packages, and cheddar cheese comes in 12-ounce packages. The American cheese is on sale for $1.59, and the cheddar cheese is on sale for $1.19. Which would you recommend Patrick buy and why?

____

____

____

____

3. In the meat department, top sirloin steak is on sale for $2.99 a pound, and rib roast is on sale for $3.09 a pound. Top sirloin steak has 162 calories and 5.8 grams of fat per 3-ounce serving. Rib roast has 198 calories and 11.1 grams of fat per 3-ounce serving. Which is the better buy? Which would be healthier for Patrick to buy?

____

____

4. This week, haddock is on sale at $2.29 a pound, and swordfish is on sale for $1.19 for 8 ounces (there are 16 ounces in 1 pound). Fried haddock has 140 calories and 5.0 grams of fat per 3-ounce serving. Swordfish has 264 calories and 8.8 grams of fat per 6-ounce serving. Which item is the best buy? Which item has fewer calories and the least fat per ounce?

____

____

Exercise 49 Application Balancing Your Diet

Name ______________________________ Date ______________

Jim and Judy are very health-conscious. They exercise regularly and eat a balanced diet. Answer the following questions.

1. Jim wants to limit his diet to 2,200 calories per day. For breakfast, he had 375 calories and for lunch he had 850 calories. To keep his calorie level to 2,200, what is the maximum number of calories he can have for dinner?

   ____

2. Judy and Jim like to jog every day. They know that they will burn about 300 calories for every 30 minutes they run. They each want to burn 2,000 calories this week. How much time will they need to spend jogging to burn the 2,000 calories?

   ____

3. Judy wants to keep to a limit of 1,500 calories per day. Look at the menu below and choose one item Judy should eliminate from each meal to help her maintain her 1,500 calorie intake.

   ____

Breakfast:	$\frac{1}{2}$ grapefruit	37 calories
	1 egg	92 calories
	1 slice toast	67 calories
	4 strips bacon	144 calories
	1 glass milk	242 calories
Lunch:	tossed salad	95 calories
	1 tablespoon dressing	67 calories
	$\frac{1}{2}$ tuna sandwich	276 calories
	1 cup cantaloupe	57 calories
	1 slice pound cake	302 calories
	diet cola	6 calories
Dinner:	6-ounce top round steak	306 calories
	20 French fried potatoes	222 calories
	1 slice French bread	85 calories
	$\frac{1}{2}$ cup lima beans	27 calories
	diet cola	6 calories

Exercise 50 Application — Spending Wisely

Name ______________________________ Date ______________

Sean works at Michael's Health Club as a fitness trainer. His appearance and health is very important in his job. Answer the following questions.

1. Sean gets a physical examination once a year to make sure that he is in good health. The cost of this annual physical examination is $375. How much money should Sean save each month so he can pay for the examination?

   ____

2. Sean goes to his dentist twice a year to have his teeth cleaned. The cost of a cleaning is $57. He also has his teeth X-rayed as part of his checkup once a year. The cost of these X rays is $60. How much does Sean spend each year on his teeth?

   ____

3. Sean takes vitamins daily. One bottle of Vitamin C will last him four weeks. How many bottles will he need to buy in one year? Each bottle costs him $6.25. How much does he spend on Vitamin C in one year?

   ____

   ____

5. Sean just got a call from his dentist. His X rays show that he has two cavities. His dentist charges $87 for each filling. He wants to pay it off in three monthly payments. How much will he have to pay each of the three months?

   ____

6. Sean is thinking about getting braces for his teeth. The dentist told him they would cost about $2,700. If he pays them off over a period of three years, what will his monthly payment be?

   ____

Exercise 51 Application Buying Clothes

Name ______________________________ Date ______________

Eric works as an assistant manager in the men's department at a clothing store. He works two nights during the week and on weekends. He makes $6.75 an hour and is paid once a week. Answer the following questions.

1. Eric needs to buy two new shirts for work. The shirts he wants to buy sell for $19.95. They are on sale now for $14.50. What will he spend on two shirts at the sale price? How much will he save buying the shirts on sale?

   ____

2. The store is having a sale on dress ties. They usually sell for $22.50 each. They are now on sale for $18 each. Eric has decided to buy four ties. How much will Eric save on each tie that he buys? What will the four ties cost him? How much will he save in all?

   ____

   ____

   ____

3. Eric needs new shoes for his job. One store has the shoes he wants on sale for $52.95. They usually sell for $69. Eric found a pair he likes at another store for $49.99. They usually sell for $59.99. How much would Eric save on the first pair of shoes? On the second pair of shoes?

   ____

4. Eric wants to buy a new suit for work. The suit he likes costs $275. If he works 30 hours each week, how many weeks will he need to work to pay for the suit?

   ____

5. How much does Eric make in one week if he works 30 hours? About $\frac{1}{3}$ of his salary is payroll deductions. How much is deducted from his check each week? How much does he get after deductions?

   ____

   ____

Exercise 52 Application Buying Sale Items

Name ______________________________ Date ______________

Jennifer usually buys clothes and other necessities when they are on sale. Use the ad below to answer the questions.

Miles Stockwell's
Aloe Skin Lotion
16-ounce body lotion

$3.99 each
(Regularly $4.99)

Not available at all locations...Limited to stock on hand.

1. How much will Jennifer save if she buys this lotion on sale? She wants to buy three bottles altogether since they are on sale. How much will she spend on three bottles? How much will she save buying three bottles?

____

2. Why would it be a problem if she plans to buy three bottles?

____

____

3. Jennifer also has to pay 5% sales tax. How much tax will she pay for one bottle? How much will she pay altogether including tax for the one bottle of lotion?

____

4. There are two limitations listed in this advertisement. What are they and what do they mean?

____

____

5. What is the cost (to the nearest cent) of one ounce of the lotion on sale?

____

Exercise 53 Analysis Reading the Ads

Name ____________________ Date __________

Read the ad. Then answer the following questions.

1. Can you determine what the original price of the television was? If so, how much was it before it went on sale?

   ____

2. What does the * after "only $59/month" mean?

   ____

3. Is there a charge to have the television sent to your home? If so, how much is the charge for this service?

   ____

4. If you had the Smith Brothers Charge Card and you were to pay only $59 per month, how many months would it take to pay for the television? (Assume there is no finance charge.)

   ____

5. The current tax rate is 8%. How much tax will you pay for this television? What will the total price be including tax?

   ____

Exercise 54 Application — Organizing a Sale

Name ______________________ Date ____________

A. Jeri is the manager of a women's clothing store. Next week, all of the summer merchandise will be going on sale. She has to decide the kind of ads she wants to run in the local newspaper to help advertise the sale. Answer the following questions.

1. Jeri purchased swimsuits for $21 each. She marked them up to $43. How much profit would she make on each swimsuit?

   ____

2. Jeri wants to mark all swimwear down 30%. How much will she mark down a swimsuit that sells for $43? What will the sale price be?

   ____

3. Jeri purchased ladies' shorts for $7.50 a pair. She marked them up 80%. What was the selling price for a pair of shorts? How much profit did she make on each pair of shorts?

   ____

4. Jeri has decided to mark the shorts down 30%. How much will a pair of shorts cost on sale? How much profit will she now make on a pair of shorts?

   ____

B. Jeri is writing an advertisement to put in the newspaper. List four other items Jeri should include in this ad.

Sizzling Summer Sale!
All Swimwear—40% off
All Summer Clothes on Sale

Limited to stock on hand

______________________ ______________________

______________________ ______________________

Exercise 55 Application Finding Bargains

Name ______________________________ Date ______________

Linda will begin her job as a legal secretary in two weeks. She needs to purchase clothing and accessories that are appropriate for a law office. Answer the following questions.

1. The local department store has women's suits on sale for 25% off. They regularly sell for $119. The catalog outlet store sells women's suits for $99. Which would be most reasonable for Linda, the local department store suits or the catalog outlet store suits?

   ____

2. Linda can buy shoes through a catalog for $45 a pair, or she can buy them direct from a factory outlet store. The factory outlet store sells all shoes for 40% off. The shoes she would like originally sell for $90 a pair. With a 40% discount, how much will she pay for shoes at the factory outlet store? Would it be more reasonable to buy shoes by catalog or through the factory outlet store?

   ____

3. Linda would like to buy accessories for her new suits. The local discount store is having a sale—30% off all handbags in stock. There are two that Linda would like to buy. One sells originally for $38, and the other sells for $44. With 30% off, how much would Linda pay for each purse? If she can buy the exact same purses for $30 each by ordering through a catalog, which would be the most reasonable for Linda?

   ____

4. Linda needs a new coat. She can buy through a factory outlet that offers 35% off all coats. The local department store is having a storewide clearance sale. They are offering 40% off all women's coats. The factory outlet has a coat that regularly sells for $200. The local department store has the same coat marked at $220. Figure the discounts at each store and decide which would be the best buy.

   ____

   ____

Exercise 56 Application — Comparing Discount Stores

Name ______________________ Date __________

Michael's Discount Wearhouse sells major brands at reduced prices. They mark down their merchandise from 30% to 70% off the retail price. Answer the following questions.

1. Erica wanted to buy a new pair of fashion designer boots. The retail price on these boots is $225. The sale ticket on the boots says they are marked down 45%, but the sale price was not on the boots. How much should Erica expect to pay for the boots?

   ____

2. Marsha has found a new sweater that is marked 50% off. The retail price on the sweater is $89. The sale price ticket says $69. Why can't this be correct? How much should Marsha expect to pay for the sweater?

   ____

3. Patricia saw a pair of jeans at a local department store for $69. The same pair of jeans at Michael's is marked at 40% off. How much will Patricia save buying these jeans at Michael's? How much will she pay for the jeans?

   ____

4. Michael's ran an ad in the newspaper saying that all men's shorts and t-shirts will be on sale for 60% off for one week only. During this week, Ryan decided to buy two pairs of shorts and three t-shirts. The shorts retail for $18, and the t-shirts retail for $12. How much will Ryan pay altogether for two pairs of shorts and three t-shirts?

   ____

5. Amy has found a swimsuit she would like to buy. She knows that the swimsuit sells for $66 at the local department store. Michael's has marked down the swimsuit 35%. How much will Amy save buying the swimsuit at Michael's? How much will she pay for the swimsuit?

   ____

6. Jerry found a jogging suit that retails for $78. Michael's sells all their jogging suits for 40% off. How much will Jerry save buying the jogging suit on sale? How much will he pay for the jogging suit at Michael's?

   ____

Exercise 57 Analysis — Ordering from Catalogs

Name ______________________________ Date ______________

Use this catalog page to complete the order form below.

Winston Smith's Merchandise

Men's jeans: 100% cotton, prewashed, preshrunk
Sizes: 30, 32, 34, 36, 38, 40, 42, 44

12303 Blue $27.00
12304 Black $29.00
12305 White $30.00

Men's short sleeve shirts:
Cotton blend, button-down collar, machine wash, front pocket
89223 White, yellow, pink, blue, green $18.00

Please add for shipping and handling: Up to $10.00—Add $2.95.
Add $1.00 shipping and handling for each additional $10.00 purchase.
California residents please add 8% sales tax.

Name Gary Welter
Address 3205 Martin Lane
City, State, Zip Los Angeles, California 90005

Catalog #	Description	How many?	Unit price	Total price
12303	Jeans—Blue	2	$27.00	________
12304	Jeans—Black	2	________	________
12305	________	1	________	________
________	Short sleeve—yellow	1	________	________
________	Short sleeve—white	2	________	________

Subtotal: ________

California residents add sales tax: ________

Shipping and handling: ________

Total Enclosed: ________

Exercise 58 Analysis — Adding Shipping and Handling

Name ______________________ Date __________

Scott Wilson Catalog Sales uses the following chart to determine how much shipping and handling to charge its customers. Use the chart below to answer the following questions.

Shipping and Handling	
0–$15.00	Add $2.50
$15.01–$25.00	Add 12%
$25.01–$50.00	Add 10%
$50.01–$100.00	Add 8%
over $100.00	Add 5%

1. In the Wilson's catalog, a pair of hiking boots sell for $49.95. Since the boots are going to be shipped to California, Wilson's must also charge 8% sales tax. Find the total a customer in California would pay to have a pair of hiking boots shipped, including tax.

   ____

2. Jason has found a fishing rod for $37.95 and a tackle box for $19.95 in the catalog. How much will Jason pay for the fishing rod and the tackle box altogether including shipping and handling?

   ____

3. Janet has decided to buy her boyfriend a down parka and a pair of gloves. In the Wilson's catalog, the down parka sells for $129.95, and the gloves sell for $27.95. How much will Janet pay for these two items altogether including shipping and handling?

   ____

4. Dustin is a great skier. In the catalog, he has found three items for his next ski trip. The ski gloves sell for $29.95. The ski hat sells for $18.95, and the ski goggles sell for $26.95. How much will Dustin pay altogether including shipping and handling charges?

   ____

Exercise 59 Application Hunting for Bargains

Name ______________________ Date __________

Ace Department Store is having a big sale this week. Answer the following questions.

1. The men's department is having a sale on suits and sports jackets. All suits are 33% off, and all sports jackets are 40% off. Find the cost of a suit originally priced at $225 and a sports jacket originally priced at $119. There is 8% tax on these items. Find the total cost including tax on the suit and the sports jacket.

   ____

2. The women's department is having a sale on misses jeans and winter coats. All jeans are 30% off, and all winter coats are 35% off. How much would you save on a winter coat that originally sells for $175? How much would you save on two pairs of misses jeans that originally sell for $45 each? What is the total amount you would save on the winter coat and the two pairs of misses jeans altogether?

   ____

3. The small appliance department is having a sale on all blenders and coffeemakers. A blender originally sells for $38, and a coffeemaker originally sells for $59. Both items are on sale for 25% off. What would the total cost of the two items be on sale including 8% sales tax? Michael's charges 5% to have the items delivered. Find the total cost including tax and delivery on both items at the sale price.

   ____

4. The children's department is having a clearance sale. All items on the clearance rack will be marked down an additional 25% at the cash register. An infant's jumpsuit originally sold for $38. It is marked down 30%, but the actual price cannot be read. How much should the actual price tag be marked? If you were to take this item to the register, how much more will be deducted from the price? What is the final sale price?

   ____

5. In the young men's department, socks are on sale for $1.29 a pair. A package of six costs $7.14. Which is the better buy, the package of six or the single pair?

   ____

Exercise 60 Application **Buying a Car**

Name ______________________ Date ____________

Tony has a new job as an electrician for a local construction firm. He will be paid $13.35 an hour and will work a 40-hour week. He needs to buy a car so he can drive to work. Answer the following questions.

1. Tony can spend about 25% of his gross monthly salary on his car payments and insurance. How much will he be able to spend on his car payment and insurance? (Hint: There are four weeks in one month.)

   ____

2. Tony knows that car insurance will cost him $125 each month. How much will he have left to spend on car payments?

   ____

3. The first car lot Tony goes to has a convertible on sale for $15,000. If he pays it off in three years, what will his monthly payment be, not including interest? Is this car within his price range?

   ____

4. The second car lot that Tony goes to has a four-door sedan for $14,500. If he pays it off in four years, what will his monthly payment be not including interest? Is this car within his price range?

   ____

5. The third car lot Tony goes to has a minivan on sale for $16,600. If he pays it off in five years, what will his monthly payment be not including interest? Is this car within his price range? Why would this vehicle be the best of the three for Tony?

   ____

6. The fourth car lot Tony goes to has a luxury sedan on sale for $22,750. What would Tony's monthly payment be if he chose to pay it off in six years? Would you recommend Tony purchase this vehicle at this time? Why or why not?

   ____

   ____

Exercise 61 Analysis Using a Blue Book

Name ______________________ Date ____________

Study the information from the blue book. Then answer the following questions.

1989 Ford Body type	VIN	Wt.	List	Whls.	Sugg. Ret.
1989 Ford—1F2A(N02C)-D-# Mustang—4-Cyl.—Equipment Schedule D W.B. 96.6′, 97.7" (2D); CID 87					
2.2 Fastback	N02C	1,927	9,890	2,500	4,600
5.0 Fastback	N22C	2,112	11,565	3,650	5,000
2.2 Convertible	D02B	2,125	10,975	2,890	4,800
5.0 Convertible GT	D22B	2,210	13,860	4,100	6,200

1. What is the wholesale price of the 2.2 convertible Ford Mustang?

   ____

2. Which of the four cars has the highest list price? Which has the lowest list price?

   ____

3. Which vehicle has the highest suggested retail price? ______________________

4. What is the difference between the wholesale and the suggested retail price of the 5.0 fastback? ______________________

5. Which of the following vehicles has a higher suggested retail price—the 2.2 fastback or the 2.2 convertible? How much more? ______________________

6. A California car dealer is selling the 1989 5.0 convertible for $6,500. Is this a good deal? Why or why not?

   ____

   ____

7. The current tax rate in California is 8.25%. Find the total cost including tax on the vehicle in problem 6. ______________________

Exercise 62 Application — Making a Down Payment

Name ______________________________ Date ______________

Larry needs to think about the amount of the down payment for a new car. He only has $3,000 in the bank. He wants to be sure he has money left over after he makes the down payment. Larry has decided he will only spend a total of $2,000 on a down payment so that he leaves some money in the bank. Answer the following questions.

1. In the newspaper, Larry has found a used station wagon for $8,000. He can finance it through his bank if he comes up with a 25% down payment. How much of a down payment on this station wagon will the bank require? Is this the amount Larry wants to spend?

   ____

2. On television, Larry saw a used convertible for $9,500. The car dealer will finance the car if Larry puts 20% down. Can he afford to do this without going over the limit he has set for himself? How much is the dealer asking Larry to put down on the car?

   ____

3. Larry's friend wants to sell a van for $3,950. Larry can take over the existing loan if he will pay 45% down. Can Larry afford to do this and still maintain the balance he wants in his account? How much money does Larry's friend want for a down payment?

   ____

4. Larry has found a new four-door sedan for $13,000. The down payment required to purchase the vehicle is $\frac{1}{4}$ of the selling price. How much will he be expected to put down on this vehicle? Can he afford this?

   ____

5. How much money does Larry want to still have in the bank after he makes a down payment on a car?

   ____

6. Which would be less—25% down on $9,500 or $\frac{1}{3}$ down on $8,100?

   ____

Exercise 63 Analysis — Paying Car Loans

Name ______________________________ Date ______________

Study the chart below. Then answer the following questions.

Rate	Number of Months to Pay Back Loan		
	12 months	24 months	36 months
10%	175.83	92.29	64.53
11%	176.76	93.22	65.48
12%	177.70	94.15	66.43
14%	179.57	96.03	68.36
16%	181.46	97.93	70.31
18%	183.36	99.85	72.30
20%	185.27	101.79	74.33
22%	187.19	103.76	76.38

1. Jordan borrowed $2,000 for 24 months. What would his monthly payment be if he was paying 14%? How much would he save each month if he could get the loan for 10%?

____

2. Dennis borrowed $2,000 for 36 months. What would his monthly payment be if he was paying 12% interest? What would his monthly payment be if he was going to borrow the same amount at the same interest level for 24 months instead of 36 months? How much would he save if he were to pay off the loan in 24 months instead of 36 months?

____

3. Andrea wants to borrow $2,000 to make a down payment on a used car. Bank A will loan her this amount at 18% interest for 24 months. Bank B will also loan her the same amount, but they will charge her 16% interest for 36 months. Which bank will charge her more interest for this loan, Bank A or Bank B?

____

4. Oscar already has $1,500 he can use for a down payment on a new car. He needs to borrow $2,000 more so that he can make a total down payment of $3,500. His credit union will loan him this amount if he pays it back in 24 months at an interest rate of 12%. If he borrows the $2,000 from the credit union, how much will he pay back to the credit union altogether?

____

Exercise 64 Application — Making a Commission

Name ______________________________ Date ______________

Eric works full-time as a salesman at a used car dealership. He receives an hourly salary plus 3% commission based on what he sells. He gets paid $400 per week. Answer the following questions.

1. On Monday, Eric sold two cars. He sold one car for $7,500, and he sold another car for $6,000. How much commission did he make on the two sales?

   ____

2. On Tuesday, Eric sold a car for $9,000. The wholesale price of the car was $6,500. How much commission did Eric make on this sale? How much profit did the dealership make on this sale after paying Eric's commission?

   ____

3. On Wednesday, Eric sold a car for $7,750. His customer put down 25% on the car. How much cash did the customer put down on the car? How much was the customer financing on the car? How much commission did Eric make on this sale?

   ____

4. On Thursday, Eric sold two cars. He sold one for $5,200, and he sold another car for $4,400. How much commission did he make altogether on both cars?

   ____

5. On Friday, Eric sold a car for $7,200. The customer put 20% down and financed the balance. How much did the customer put down in cash on the car? How much was financed? How much commission did Eric make on this sale?

   ____

6. On Saturday, Eric had a difficult time making a sale. He only made one sale of $2,600. How much commission did he make on this sale?

   ____

7. Using the information from problems 1 through 6, how much commission did Eric earn for the week? Eric can earn a bonus of $100 if his total sales for the week are at least $25,000. Find Eric's total income for the week including salary, commissions, and bonus.

   ____

Exercise 65 Application Comparing Auto Insurance

Name ______________________________ Date ______________

Melinda just bought a new convertible. She will make 48 monthly payments of $127.50 in addition to the $2,500 she put down on the car. The bank that approved her auto loan has to have proof of insurance from Melinda. Answer the following questions.

1. Smith's Auto Insurance charges $1,200 per year for the kind of policy Melinda wants for her new car. What will Melinda's monthly payment be if she divides it evenly over 12 months? What will her total monthly expenses be on the car and its insurance?

   ____

2. A.B.C. Insurance charges $1,350 per year for the same policy, except that they give a 10% discount to drivers with no tickets on their record. How much will Melinda save per year on the A.B.C. policy if she has no tickets? What will her monthly payment be on this policy?

   ____

3. Northwest Insurance has a ten-pay plan instead of monthly payments on its insurance policies. For the same policy that Smith's and A.B.C. offer, they would charge Melinda $119 per payment on the policy she wants. What will the total annual cost of this policy be?

   ____

4. Southern States Insurance Company has a similar policy specifically for new drivers. They have a six-month policy instead of a yearly policy. The six-month policy that Melinda is interested in costs $525 altogether. If she was to be allowed to pay it in monthly installments, how much would she pay each of the six months?

   ____

5. Insurance Trust offers the same policy as the other auto insurance companies. They require each insured person to pay 20% down on the policy. The balance can then be paid off in 10 installments during the year. Their policy costs $1,100 altogether. How much would Melinda have to put down on this policy? What would her monthly payments be after she makes the down payment?

   ____

6. Which of the insurance companies listed above offers the best rate for Melinda?

   ____

Exercise 66 Application — Applying for Insurance

Name ______________________________ Date ______________

Study the driving record. Then answer the questions below.

Surething Insurance Company

Applicant: Carol Lee Johnson
1432 Erickson Road
Brockport, NY 14420

Listed below are items on your driving record that are part of the rating of this policy. Please contact your agent if there are any errors in this report.

Carol Lee Johnson

Conviction/Accident	Date	Points
Driving Record Clean	08/04/90	00

Rhonda Sue Johnson

Conviction/Accident	Date	Points
At-fault accident	08/27/88	04
Moving Violation	11/30/87	01

1. Suppose that the basic rate for the insurance Carol Lee wants is $658 per year. Suppose that a total of five points will add 40% to that insurance rate. How much will the insurance cost?

 ____

2. Suppose it takes three years for an accident to be erased from a record. When will Rhonda Sue's record be clean?

 ____

3. How many months passed between the time Rhonda Sue got a ticket and the accident?

 ____

4. By the time Rhonda Sue's record is clean, the basic rates have gone down. Because the car is older, it costs less to insure it. The basic rate is now $521. How much less is this than the amount they were paying?

 ____

Exercise 67 Application — Insuring a Vehicle

Name ______________________________ Date ______________

Dustin needs to purchase auto insurance for his new car. He wants to be sure that the policy will include liability, collision, comprehensive, and uninsured motorists. Several insurance companies have given him rates on the types of coverage that he wants. Answer the following questions.

1. Dustin has several options of liability for his auto insurance policy. He has a choice of the following:

   15/30 for $169 per six months

   50/100 for $199 per six months

   100/300 for $224 per six months

   300/300 for $280 per six months.

   Which coverage gives him the most liability coverage? What is the monthly charge for that coverage?

   ____

2. Dustin wants to purchase Property Damage Coverage. He can purchase $5,000 per accident for $41 per six months, $10,000 per accident for $44 per six months, $25,000 per accident for $47 per six months or $50,000 per accident for $50 per six months. Which type of coverage would you recommend for Dustin and why?

   ____

3. Dustin can purchase Medical Coverage for his new car. His options are $1,000 each person for $31 per six months, $2,000 each person for $38 per six months or $5,000 each person for $46 per six months. What would the difference in monthly payments be between the highest and lowest coverage?

   ____

4. Collision is the most expensive coverage for a new car. With a $100 deductible, Dustin will pay $511 for six months. A $200 deductible will cost $402 for six months. A $300 deductible will cost $343 for six months. A $500 deductible will cost $309 for six months. What is the monthly charge for a $100 deductible? What is the monthly charge for a $500 deductible? Which would you recommend for a new car and why?

   ____

   ____

Exercise 68 Analysis — Buying Tires

Name ______________________ Date __________

Study the chart. Then answer the following questions.

Kind of tire	How long it will last	Price
Radial	about 40,000 miles	$90.00
Belted bias	about 25,000 miles	$75.00
Bias ply	about 15,000—20,000 miles	$65.00

1. Ken noticed that his front tires were beginning to show signs of wear. He already had 40,000 miles on his car. He wanted to sell his car once he had 60,000 miles on the car. Which of the above type tires would you recommend he buy and why?

____

____

2. Clay noticed that the spare tire in his trunk was flat. He decided to replace it with a new tire. Since it's just a spare, Clay knows it doesn't have to be the best tire. The other four tires on his car were all radials. At the current prices listed above, which tire would you recommend Clay purchase and why?

____

____

3. Evan needs to purchase a complete set of tires for his car. He has decided to purchase the least expensive tires since his car is not in the best shape. He must also pay 7% sales tax. How much will he pay altogether for the tires including tax?

____

4. Angela wants to purchase two tires since they are on sale. She has radial tires on her car now, so she wants to buy the same type of tire. They are on sale for 25% off. How much will Angela spend on the two tires altogether with the discount and 8% sales tax?

____

5. Nicholas can't decide whether to purchase a radial or a belted bias for his car. The radial is on sale for 25% off, and the belted bias is on sale for 40% off. Which tire will be the least expensive at these sale prices?

____

Exercise 69 Application Repairing Vehicles

Name ______________________________ Date ______________

Clint works full-time as an auto mechanic at a service station. He works 40 hours per week and is paid $12.25 an hour. He gets paid every other week and has 28% deducted from his check. Answer the following questions.

1. On Monday, Clint worked on a pickup truck that needed a tune-up and an oil change. The parts came to a total of $17.50, and the charge for labor was $97. There is a charge of 7% tax on the parts only. What are the total charges?

   ____

2. On Tuesday, Clint serviced three different cars. On the first car, he did a front-end alignment and rotated the tires for a total of $69. On the second car, he flushed out the carburetor and put in a new air filter for a total of $96. On the third car, he flushed the radiator and put new coolant in for a total of $49. Find the total charges for the three vehicles Clint worked on Tuesday.

   ____

3. On Wednesday, Clint worked on a car that was towed in with a dead battery. The cost of the battery was $87. He charged the customer $26 for putting the battery in. There is sales tax of 7% on the parts but not on the labor. There was also a charge of $50 for towing. Find the total cost of the bill.

   ____

4. On Thursday, Clint had to work 11 hours since there was so much repair work to be done. He repaired 3 cars and 2 trucks for a total charge of $477.40. What were Clint's gross wages (before deductions) for the day? About how much profit did the service station make on Clint's work?

   ____

5. On Friday, Clint repaired a minivan with a transmission problem for $241. He also repaired a compact car with an oil leak for $89. Finally, he repaired a pickup truck with a faulty fuel pump for $138. What was the average cost per job?

   ____

Exercise 70 Application — Solving Maintenance Problems

Name ______________________ Date __________

Joel bought a used car for $12,000. He put $3,000 down and is paying off the rest over a four-year period. The warranty will cover most of the work the car might need in the next year. However, some items may not be covered by the warranty. Answer the following questions.

1. Joel's insurance policy is for a six-month period. For the six-month policy, he is paying $169 for liability, $41 for property damage, $31 for medical, $22 for uninsured motorists, $156 for collision, and $189 for comprehensive coverage. What will Joel's monthly payment be for this insurance policy? How much will he pay altogether for one year?

   ____

2. After Joel makes the down payment, how much will he still have left to pay on the car? If he were not paying interest, what would his monthly payment be? (Remember, there are 12 months in one year.)

   ____

3. The first month he owned the car, Joel noticed that his rear tires were showing signs of wear. Radial tires for his car sell for $89 each, and belted bias tires for his car sell for $158 a pair. Which would be the better buy? Using the price of the better buy, find the cost of two tires for Joel's car including tax at 8% and an $11 charge to balance and mount each tire.

   ____

4. Joel decided to take the car in for a tuneup and timing adjustment. The parts cost $37, and the charge for labor was $88. There is an 8% sales tax on the parts. What is the total bill?

   ____

5. Joel has been offered $8,500 for his car after he's had it for one year. Will he make a profit if he sells it after having it for only one year? (Hint: First you will have to figure out how much money Joel still owes on the car and compare it to the price he was offered.)

   ____

6. Joel was paying 9.5% interest on his car loan. How much interest would he pay in one year on this car?

   ____

Exercise 71 Application — Figuring Credit Card Interest

Name ______________________________ Date ______________

Jennifer has both a Charge-it Card and an Americard. The annual interest rate on each of the cards is 18%. Interest is charged only on the unpaid balance at the end of each month. Answer the following questions.

1. The annual interest rate paid for the Charge-it Card is 18%. How much interest is paid per month on the Charge-it Card? (Hint: There are 12 months in one year.)

   ____

2. When Jennifer received her Charge-it Card bill, the balance was $417. Her minimum payment due is $20. She has decided to pay $67 this month on the bill. How much interest will be added to her bill after her payment of $67 has been credited to her account? What will the new balance on this account be?

   ____

3. Jennifer's Americard account has a balance of $666. The minimum payment due is $25. She has decided to pay $50 on this account. What will the new balance on her account be after she has been credited with the payment she is making and they have added the new interest?

   ____

4. Jennifer has decided to buy a new suit for work. The suit costs $87. If she charges this to her Charge-it Card, how much will be added to her account including 8% sales tax? What will the new balance on her Charge-it Card account be? (See problem 2.)

   ____

5. Jennifer ran short of funds this month and decided to get a $200 cash advance on her Americard account. There is a 3% service charge for this transaction. How much will be charged to her account for the cash advance and the service charge?

   ____

6. Jennifer can avoid an interest charge if she pays off her Americard account in full within 30 days. She has decided to pay $200 on this account. How much will she have to pay on her next month's bill (including interest) to pay off the account in full?

   ____

Exercise 72 Application — Filling Out Applications

Name ______________________ Date __________

Jason has worked as a teacher for the last 36 months at the ABC Unified School District. His annual salary is $32,000. Jason has lived in the same apartment for 2 years. Jason wants to apply for a Gibb's Department Store Credit Card. Fill out the credit card application with the known information. Then, complete the application with any other information you want to add.

GIBB'S DEPARTMENT STORE CREDIT CARD APPLICATION

① TYPE OF ACCOUNT REQUESTED (CHECK ONE): ☐ INDIVIDUAL ☐ JOINT

② FIRST NAME | INITIAL | LAST NAME

HOME ADDRESS | APT. # | CITY | STATE | ZIP

③ ☐ RENT ☐ WITH PARENTS ☐ OWN ☐ BOARD | HOW LONG ? YRS. | HOME PHONE Include Area Code

DATE OF BIRTH | SOCIAL SECURITY NO. | DRIVER'S LICENSE NO. — STATE

④ PREVIOUS ADDRESS — If less than 2 years at current | CITY | STATE / ZIP

⑤ EMPLOYER | POSITION | HOW LONG ?

BUSINESS ADDRESS | BUSINESS PHONE Include Area Code

⑥ BANK — LIST BRANCH AND ADDRESS ☐ CHECKING ☐ SAVINGS

OTHER CREDIT REFERENCES | ACCOUNT NO.

Exercise 73 Application — Paying Bills

Name ______________________ Date __________

Study Brian's Charge-it Card bill. Then answer the following questions.

Previous Balance	Charges	Finance Charge	Credit/ Returns	Payments	New Balance	Past Due
$659.03	$201.47	$8.76	.00	$75.00	$794.26	.00

Account Number	Credit Limit	Payment Due Date	Minimum Payment Due	
705-29-3	$1,000	6-25-92	$80.00	To avoid additional finance charge pay in full by payment due date shown.

1. How much credit does Brian have left before he reaches the credit limit on his Charge-it Card account?

   ____

2. About what percent of the total bill is the minimum payment due?

   ____

3. What can Brian do if he wants to avoid paying a finance charge this month?

   ____

4. Were there any additional credits to Brian's account besides the payment he made? Is there enough information on this portion of the statement to determine when his payment was received?

   ____

5. How was Brian's new balance figured on this statement?

   ____

6. Is there any amount overdue on Brian's Charge-it Card account? How do you know this?

   ____

7. Brian has allowed for $100 in credit card payments in his monthly budget. Is he within his limit if this is his only credit card?

   ____

Exercise 74 Application — Buying on Credit

Name ______________________ Date ____________

Beth has a Charge-it Card and a Moore's Department Store Credit Card. Her monthly take-home (net) salary is $2,056. She currently has $4,500 in her savings account at the bank and a balance of $1,100 in her checking account. Answer the following questions.

1. Beth knows that the payments for her credit cards altogether should not exceed 20% of her net pay. Her monthly payment on the Charge-it Card is $83, and her monthly payment on the Moore's card is $46. How much is Beth spending each month on both credit cards? Is this less than 20% of her net pay for the month?

   ____

2. The interest rate on the Moore's card is 16%, and the interest rate on the Charge-it Card is 18%. Would Beth be wise to close her Moore's account and charge the balance to her Charge-it Card? Why?

   ____

3. Beth can get a Charge-it Card with a special interest rate of 14% through her bank. Should she combine her other credit cards into one account? Why?

   ____

4. Beth needs a new sofa for her living room. Moore's Department Store is having a sale on all living room furniture at 25% off. The sofa Beth likes originally sells for $675. What would the price of this sofa be at the sale price? She can either charge it on her Moore's card or finance it through her credit union at an interest rate of 15%. Which option would cost her the least? How do you know?

   ____

5. Beth wants to buy a car for $9,700. If she puts down 25% and pays off the rest, how much will her down payment be? For $388, she wants the dealer to install a CD player in the car. She can either put this charge on her Charge-it Card or add it to the balance on her car. If the credit union finances the car at an interest rate of 9.75%, which will save her the most money—the Charge-it Card or the credit union?

   ____

Exercise 75 Application — Figuring Interest

Name ______________________ Date __________

Angela has just rented a new two-bedroom apartment. The monthly rent for the apartment is $750 plus utilities. She is also hoping to buy several pieces of furniture on credit so that she can pay them off over a period of time. Angela already has a Furniture World Credit Card with a balance of less than $100. She is paying 15% interest on this account. Answer the following questions.

1. Angela wants to purchase a dinette set with four chairs. The set she likes is on sale at $33\frac{1}{3}$% off. It originally sells for $198. What is the sale price on this dinette set? If she puts the dinette set on her Furniture World credit card, how much interest per month will she pay on the set? How much in dollars and cents does this come to each month?

   ____

2. Furniture World is having a sale on bedroom sets that include a bed, a dresser, and a nightstand. The set Angela likes is marked down 20%. If the original price is $475, what is the sale price? She also found a bed on sale for $175 at Bedrooms Galore and a dresser for $249 at the Furniture Wherehouse. Which will offer her the best buy? How much annual interest will she pay if she purchases the bedroom set from Furniture World?

   ____

3. Angela needs living room furniture. She can purchase a sofa and loveseat from Furniture World for $525. She found a coffee table at the Furniture Wherehouse for $225. She can charge the coffee table on Furniture Wherehouse's credit plan at 16% interest and the sofa and loveseat from Furniture World on their credit plan. How much annual interest will Angela pay for all these items if she pays them on time?

   ____

4. Angela has found a wall unit with a mirror for $125. However, she cannot pay for it on time because her Furniture World account balance has reached its limit. She can finance it through her credit union for 14.4% interest. How much monthly interest will she pay on this item if she finances it through her credit union?

   ____

Exercise 76 Application — Applying for a Loan

Name ______________________________ Date ______________

Judy is going to the bank to get a $3,600 loan to do some major repair work on her car. She has worked as an assistant manager at a shoe store for the last ten months and makes $398 per week. She lives alone in a one-bedroom apartment and pays $425 per month rent. She has lived in this apartment for three years.

1. Judy has decided to request the loan at a fixed interest rate for three years. How much would her monthly payment be (not including interest)?

   ____

2. The bank charges 18% on consumer loans. What percent interest will Judy pay monthly on this loan if it is approved? How much will she pay in interest each month?

   ____

3. What will Judy's total expenses be including rent and loan payment with interest? How much of her salary will she have left each month for her other expenses?

   ____

4. Since Judy has only worked at the shoe store for ten months, what might a loan officer want to know about Judy before giving her a loan?

   ____

   ____

5. Based on Judy's salary and expenses, does it appear she would be able to make the monthly payments with interest? If you were the loan officer, what recommendation would you make?

   ____

   ____

Exercise 77 Application — Receiving a Credit Rating

Name ______________________________ Date ____________

Martin lives in a two-bedroom apartment and pays $675 rent per month. He has lived in his apartment for four years. He has worked at the loading docks for five years and makes $12.50 an hour. His monthly expenses, not including rent, are $550 per month. Answer the following questions.

1. Martin has a checking and savings account at his bank. There is a balance of $3,000 in his savings account and a balance of $357 in his checking account. Will these accounts help Martin get a good or bad credit rating?

   ____

2. Is Martin's employment history evidence of job stability or instability? Would this situation help or hurt Martin's credit rating? What is Martin's gross weekly income if he works 40 hours per week? What is Martin's gross monthly income (based on four weeks)?

   ____

3. Martin wants to make a loan from his bank for $4,000 to consolidate his bills. He is currently paying 20% interest on these bills. The bank charges 16% on a consumer loan. What percent of interest will he save each year? How much money would he save each year in interest? How much is that each month?

   ____

4. Circle one or more of the following that the bank will check before they decide whether or not to give Martin the loan.

   a. Martin's credit history
   b. How long he has lived at his current address
   c. How long he has worked on his job
   d. His age
   e. His checking and savings accounts
   f. The kind of car he drives
   g. His ability to repay the loan

Exercise 78 Application **Figuring Interest**

Name ______________________________ Date ______________

Matt is $12,000 in debt. He is comparing the different plans that his local banks and other lending institutions offer. He does not want to take out a loan that will take more than five years to pay back. Answer the following questions.

1. Matt's bank has two different plans available. The first plan is a fixed rate over five years. The interest rate is 12%. The second plan is a variable interest rate that begins at 8% and can vary (increase) at most one percent each year for the length of the loan. How much interest would Matt pay altogether on the loan at the fixed rate? How high can the variable rate go?

   ____

2. Matt financed his car through his credit union. He knows that they also offer good rates on consumer loans. After checking with them, he found out that they can offer him 9.5% on a loan over $10,000. How much interest would he have to pay on his loan?

   ____

3. Matt is currently paying $450 altogether on his bills. If he were to get a loan through the credit union, how much money would he save each month on his payments?

   ____

4. Matt is having trouble deciding whether he should take the credit union loan for four years or five years. How much interest altogether will he pay for the loan over five years? Over four years? How much money would he save by paying back the loan in only four years?

   ____

   ____

5. If Matt were to pay back the credit union loan in four years, would his monthly payments be more or less than they were before?

   ____

Exercise 79 Analysis Figuring Loan Fees

Name ______________________ Date __________

Most banks and lending institutions charge fees for loans in addition to interest. Lance was comparing rates and fees of several lending institutions before deciding to apply for a loan. Fill out the total payback section of the table below so he can choose a loan that is reasonable. Then answer the questions below.

Loan	Principal	Rate of Interest	Loan Fee	Length of Loan	Total Payback
A	$5,000	8.5 %	$ 75	3 years	__________
B	$6,500	7.75%	$125	4 years	__________
C	$6,000	8 %	$ 90	2 years	__________
D	$8,000	7.5 %	$120	3 years	__________
E	$7,500	8.25%	$100	4 years	__________

1. How much in interest would Lance pay altogether on Loan A over three years? How much interest would he pay per year? How much interest would he pay per month?

____

2. Which of the five loans above would most likely have the largest monthly payment? How would you know that?

____

3. What is the difference between the total payback for the loan with the highest fee and that for the loan with the lowest fee? __________

4. Which loan would most likely have a lower monthly payment—a loan paid back over three years or the same loan paid back over four years? Why would you expect that?

____

5. In Loan B, if Lance were to deduct the fees from the actual loan, how much would he actually receive from the loan? __________

6. Lance needs a minimum loan of $5,500. He can only afford to pay $210 per month on the loan. Which loan is best for Lance? __________

Exercise 80 Application — Financing Equipment

Name ______________________ Date __________

Ken owns a construction company. He has to buy several items to keep his business growing. Answer the following questions.

1. Ken wants to finance the purchase of a new truck through his bank. The sticker price on the truck was $9,500. He is adding $2,200 in extras to the vehicle. He needs to put 25% down so he can finance the balance. His bank will finance this new vehicle at 9.5% over five years. How much cash will Ken have to put down on the truck? How much will Ken be financing (not including tax)? Find the monthly payment for the truck over five years.

   ____

2. Ken needs to make a consumer loan to purchase tools and equipment for his business. He needs a total of $4,000 to make all the purchases. He belongs to a credit union, and they will make a loan in that amount over three years at 11%. How much interest will Ken pay altogether on this loan?

   ____

3. Ken bought a computer for $1,200 and a printer for $650. With an 8% sales tax, how much will this system cost Ken's business altogether? If the computer store allows him to pay it off over two years at 15% interest, how much interest will he pay altogether? What will be the monthly payment, including interest, for the computer and printer?

   ____

4. Ken buys lumber at the local lumber mill. He currently has a balance of $2,346 on account. Yesterday, he ordered more lumber totalling $644. Find the new total on his account. The lumber mill charges 18% interest annually. What percent interest is charged each month? How much interest will be added to Ken's account based on the new balance?

   ____

5. Ken charges all gasoline, maintenance, and repair work on his business vehicles to help keep track of his expenses. Last month, his balance on account was $1,388. This month, he made a payment of $200 and charges of $126. Not including interest, what would be the new balance on this account?

   ____

Exercise 81 Application — Figuring Recreation Costs

Name ______________________ Date __________

Mark and Carla are planning a birthday party for their twelve-year-old son Alan. They would like to take Alan and his friends miniature golfing or roller skating. Use the tables below to answer the questions.

Hole-in-One Golfing		Roller Skateland	
Monday through Friday:		**Monday through Thursday:**	
Adults (age 14 and over)	$5.50	Adults (age 13 and over)	$2.50
Children (age 5 to 13)	$4.00	Children (age 6 to 12)	$2.00
Children (under age 5)	Free	Children (age 5 and under)	$1.00
Saturday and Sunday:		**Friday through Sunday:**	
Adults (age 14 and over)	$7.50	Adults (age 13 and over)	$4.00
Children (age 5 to 13)	$6.00	Children (age 12 and under)	$3.00
Children (under age 5)	Free		
		Skate Rentals	$1.00

1. Alan would like to invite 10 of his friends to his party. Of his friends, 6 are 12 years old and 4 are 13 years old. How much would it cost Alan's parents to take Alan and his friends roller skating including the rental of skates if the party was to be held on a Saturday?

   ____

2. If Alan were to take the same friends miniature golfing, would it cost more or less than roller skating? How much of a difference would there be in the cost?

   ____

3. Alan has two brothers, age 6 and 14, and one sister, age 9. Since Alan's parents cannot find a sitter, they will have to go along to Alan's party. How much will it cost for his brothers and sister to go roller skating with Alan and his friends (including skate rentals)?

   ____

4. If they decide to go miniature golfing, Alan's parents will also want to golf. Everyone at the party will be served cake and ice cream at a cost of $2.00 per person. Find the total cost for Alan, his parents, his brothers and sister, and all his friends to go miniature golfing and eat cake and ice cream.

   ____

Exercise 82 Application — Finding Bargains

Name ______________________ Date __________

Kyle and Tara, both 16, like to go to the movies. They want to save as much money as possible, so they check the newspapers for special showings and bargain features. Use the movie schedule below to answer the following questions.

Fashion Cinema Theater		Mercury Theater	
Adults	$7.50	Adults	$7.00
Children (under 13)	$6.00	Children (under 13)	$6.00
Matinees (Mon.–Fri.)	$4.50	All Matinees	$5.00
Matinees (Sat.–Sun.)	$5.50	Sneak Previews	$4.00
Sneak Previews	$5.00	Senior Citizens	$3.00
Senior Citizens	$3.50		

1. Both theaters happen to be showing the same feature this week. How much would it cost Kyle and Tara if they attended a Saturday matinee at the Fashion Cinema Theater? How much would it cost them for a Saturday matinee at the Mercury Theater? How much would they both save at the Fashion Cinema if they were to attend a Friday matinee instead?

   ____

2. Kyle and Tara like to go to the movies twice a week. How much would they save altogether at the Mercury Theater if they attended matinees instead of evening showings?

   ____

3. Kyle and Tara have asked two of their friends to attend a sneak preview at the Fashion Cinema. How much will it cost them altogether if they also spend $3.50 each on refreshments?

   ____

4. This week, the Fashion Cinema is showing one film while the Mercury Theater is showing two movies for the price of one. If Kyle and Tara attend weekday matinees at both theaters, what is the average cost per movie?

   ____

Exercise 83 Evaluation — Getting the Best Deal

Name ______________________________ Date ____________

Jon and Jillian want to join a health club. Compare the membership plans for three health clubs in the chart below. Then answer the following questions.

The Spa	The Club	Top Fitness
12 months.... $169	12 months.... $188	12 months.... $135
6 months...... $ 89	6 months...... $ 99	6 months...... $ 75
Family pass... $ 25/month	Family pass... $ 30/month	Family pass... $ 35/month
Daily pass..... $ 5	Daily pass..... $ 4	Daily pass..... $ 5
Weekly pass... $ 12	Weekly pass... $ 10	Weekly pass... $ 11

1. Jon and Jillian have never belonged to a health club before. They have decided to try a weekly pass at each location to see which of the three clubs they like best. How much will it cost altogether for both Jon and Jillian to purchase a weekly pass to each of the three clubs?

____

2. They are also considering a one-year membership. Which club offers the best annual membership fee?

____

3. How much would they save at The Club if they purchased annual memberships instead of six-month memberships for the both of them?

____

4. Jon and Jillian plan to work out three times a week. If they choose Top Fitness, would they save more money buying daily or weekly passes? How much would they save per person?

____

5. The Club is two blocks from Jon and Jillian's home. The Spa is 8 miles from their home. Working out three times each week, they figure they will spend $8 each month on gas if they join The Spa. Would it be less expensive for both Jon and Jillian to join The Spa or The Club (include the cost of gas) for one year? How much difference would there be in the costs?

____

Exercise 84 Application Shopping for Sporting Goods

Name ____________________ Date ____________

Jerry and Susan love to camp. They have been watching for sales at the local sporting goods store. Answer the following questions.

Nicky's Sporting Goods

Pup tent	$27.95	Canteen	$ 8.95	Ice chest	$14.95
Tackle box	$14.88	Barbeque	$18.79	Flashlight	$ 6.75
Sleeping bag	$31.59	Backpack	$17.09	Vest	$18.47
Folding table	$22.49	Mess kit	$ 4.99	Thermos	$ 8.99

1. Nicky's Sporting Goods is having a sale. They are marking all camping items 25% off. Jerry and Susan need two new sleeping bags and two new backpacks. How much will these items cost altogether on sale? They also have to pay 7% sales tax. What will the total cost be including tax?

   ____

2. Another sporting goods store sells pup tents for $22.95. With 25% off the pup tent at Nicky's, will it cost more or less than at the other sporting goods store? Find the difference in these prices. How much will Jerry and Susan spend if they purchase the less-expensive tent? (Don't forget to add 7% sales tax.)

   ____

3. Jerry and Susan also both need vests. In addition to the 25% off that is being advertised, Nicky's is also taking an additional 10% off the reduced price for a one-day sale. How much altogether will be discounted from each vest? Find the cost of two vests including tax during the one-day sale.

   ____

4. Jerry and Susan also need a new ice chest, one canteen, and two mess kits. Including tax, find the total cost of these items on sale.

   ____

5. Which item listed above will show the largest dollar savings? How much would be saved if the item was purchased on sale? Find the cost including tax on this item.

   ____

Exercise 85 Application — Solving Budget Problems

Name ______________________________ Date ______________

Eric and Lisa are going camping for two weeks in the mountains. They are making all the necessary plans and buying everything they will need. Answer the following questions.

1. The round trip to the mountains is about 600 miles. Their station wagon gets 24 miles to the gallon of gas. The current price for gas is $1.39 per gallon. How much money should they allow in their budget for gasoline?

   ____

2. Eric and Lisa have figured they will need to buy enough food for three meals a day for two weeks. How many meals will each consume during this period of time? They have determined that the average cost per meal is about $3. How much money should they budget for food for the two of them?

   ____

3. Sports World is having a sale on camping equipment this week. Backpacks are on sale for $23.99, and sleeping bags are on sale for $27.95. The current sales tax is 8%. How much will they spend altogether including tax if they purchase two backpacks and two sleeping bags?

   ____

4. The Parks Department rents campsites by the week for $49. Eric and Lisa must pay for the campsite at least three weeks in advance. If they do not make the payment in advance, there is a 10% penalty. How much will they pay for the two weeks if they are late sending a check? How much penalty would they have to pay?

   ____

5. To be safe, they decide to have their car serviced before they go on their trip. The car gets a tune-up and an oil change. The cost of the parts is $23.95, and the charge for labor is $87. There is 8% sales tax on the parts only. How much is the total cost of the work done to their car?

   ____

6. Find the total cost of the trip using the information from problems 1 through 5. (They have paid for their campsite in advance.)

   ____

Exercise 86 Application — Planning a Trip

Name ______________________________ Date ____________

Frank and Jessica want to visit the northwest region of the United States by car. They are now in the planning stages of the trip and need to figure out costs, time factors, and mileage. Answer the following questions.

1. According to the map they have received from their auto club, the total distance Frank and Jessica will travel is approximately 3,675 miles. How many miles will they need to travel each day to complete the trip in three weeks? (Hint: There are seven days in one week.)

   ____

2. They have determined that their average speed will be 50 miles per hour while they are on the road. About how many hours should they plan on driving the car each day? (Round your answer to the nearest whole number.)

   ____

3. Frank and Jessica have figured that the average cost per night, per person is about $23. How much should they plan on spending on hotel/motel accommodations for the entire trip?

   ____

4. The family car gets about 25 miles to the gallon. How many gallons of gas will it take to complete the trip? Gasoline costs about $1.49 per gallon. How much will it cost Frank and Jessica for gas for their trip?

   ____

5. Frank and Jessica only plan on eating two meals each day while on their vacation. They figure that the average breakfast will cost $3.99 per person, and the average dinner will cost $6.99 per person. How much should they plan on spending on food for the entire trip?

   ____

6. Not including souvenirs and other expenses, how much should Frank and Jessica plan on spending for their trip? (Include cost of gas, food, and lodging only.)

   ____

Exercise 87 Application — Taking a Flight

Name ______________________________ Date ______________

Study the flight information in the chart. Use the chart to help you measure time.

Flight Information

Flight Number	Destination	Time of Departure	Time of Arrival
12	Newark, New Jersey	7:30 A.M.	4:35 P.M.
22	Newark, New Jersey	10:00 A.M.	9:50 P.M.
34	Newark, New Jersey	11:40 A.M.	8:45 P.M.
67	Newark, New Jersey	5:20 P.M.	4:25 A.M.

All of these flights leave from Los Angeles, California.

1. Flight 12 goes directly from Los Angeles to Newark without stopping. How much time does the flight appear to take?

   ____

2. There is a three-hour time difference between Los Angeles and Newark. If it is 4:35 P.M. in Newark, it is 1:35 P.M. in Los Angeles. How much time does Flight 12 really take?

   ____

3. Flight 22 is not a direct flight. It stops in Dallas to pick up more passengers. How much longer does Flight 22 take than Flight 12?

   ____

4. How much time does Flight 67 take? (Remember that there is a three-hour time difference.) Is Flight 67 a direct flight?

   ____

5. Suppose a direct flight left Newark at noon. It is a six-hour flight. About what time would it arrive in Los Angeles? (Remember it will be three hours earlier in Los Angeles.)

   ____

Exercise 88 Application Finding Lodging and Food

Name ______________________________ Date ____________

Sergio and his family are planning a two-week vacation to the Great Lakes. Answer the following questions.

The Great Lakes

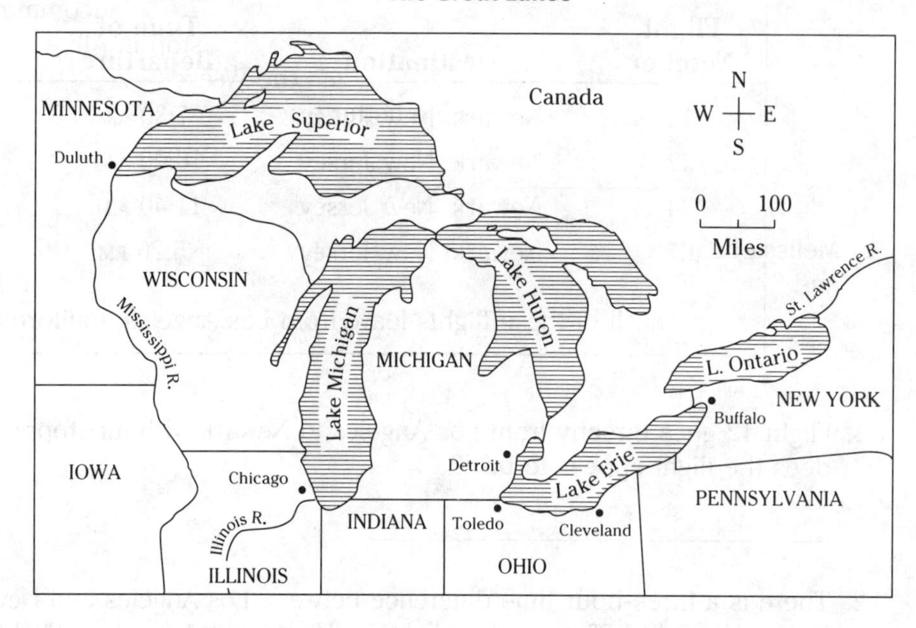

1. Sergio and his family will fly from Denver to Chicago, a distance of 1012 miles. The plane flies at an approximate air speed of 500 miles per hour. Estimate the actual flying time one way from Denver to Chicago. How many hours should it take to make this trip? How many minutes does this equal? ____________

2. Once in Chicago, they are going to rent a car and drive north through Wisconsin. The actual distance is about 500 miles. If Sergio drives at an average speed of 40 miles per hour, how long would it take to make this trip if he were to make no stops?

   ____

3. The trip from Milwaukee to Mackinac Island takes about 6 hours and 45 minutes. The trip from Mackinac Island to Traverse City, Michigan takes about 8 hours and 34 minutes, and the trip from Traverse City to Chicago takes about 4 hours and 52 minutes. What is the total time spent traveling from Milwaukee to Mackinac Island to Traverse City and then to Chicago? (Write your answer in hours and minutes.)

   ____

4. The flight from Chicago to Denver does not take as long as the flight from Denver to Chicago since there is a tailwind. The trip takes about 20 minutes less altogether.

   How long is the flight from Chicago to Denver? ____________

Exercise 89 Application — Figuring Gas Mileage

Name ____________________ Date __________

Melissa and Kurt are planning a three-week car trip through the Southwest. They plan on driving about 250 miles each day so they can cover the greatest area and still have time to sightsee. Answer the following questions.

1. Melissa and Kurt have taken the time to check out all the accommodations available along the way and have decided to make their first stop in Albuquerque. The Day/Nite Motel charges $34 per room per night, and the Weekender Lodge charges $18 per person per night. Which offers the better price? By how much per night?

   ____

2. Melissa has figured that they will spend about $4 each on breakfast, $5 each on lunch, and $8 each on dinner. How much will be spent by each per day on meals? What should they expect their total food bill to be for the two of them for the entire three-week trip? (Hint: There are seven days in one week.)

   ____

3. A friend of Kurt's has a discount coupon for the Highlands Hotel in Houston. The regular rate is $27 per person per night. The coupon gives them 25% off the regular price. The Resorts Inn has a regular rate of $37 per night per room. Which will offer them the best price? How much will they save staying in the less expensive room?

   ____

4. In New Orleans, they can stay in the French Quarter for $54 per room per night. However, at this time of year there is a special rate that allows a 15% discount on each night after the first night's stay. If they should decide to stay three nights, how much will the room cost them altogether?

   ____

5. On the return drive home, they decide to stop at a fast-food restaurant. Kurt ordered a hamburger for $1.99, french fries for $1.29, and a soft drink for $.99. Melissa ordered a chicken sandwich for $2.09 and a soft drink for $.79. The current sales tax rate is 6%. Including tax, how much did they spend altogether on this lunch? (Round your answer to the nearest cent.)

   ____

Exercise 90 Application — Figuring a Budget

Name ______________________ Date __________

Beth and Paul are planning a road trip for their two-week summer vacation. Answer the following questions.

1. When the gas tank of their car is full it holds 20 gallons of gas. The cost of gas locally is $1.39 per gallon. How much would it cost them to fill the tank if it was completely empty?

   ____

2. On the first leg of their journey, they traveled 250 miles from Los Angeles to Phoenix. They used exactly $\frac{5}{8}$ of a tank for this part of the trip. How many miles per gallon did they average on this part of the trip?

   ____

3. The second leg of the journey was all freeway driving. They traveled the next 300 miles on $\frac{4}{8}$ of a tank of gas. How many miles per gallon did they average on this part of their trip?

   ____

4. Compare the answers from problem 2 and problem 3. Give at least one reason why the mileage rates can differ so much with the same car.

   ____

5. On the last day, Beth and Paul averaged 24 miles per gallon and used $\frac{3}{5}$ of a tank of gas. How many miles did they travel on this part of their trip?

   ____

6. Beth and Paul decided to make the return trip in half the time of the original trip. They used a complete tank of gas and averaged 26 miles per gallon. How many miles did they travel on their return trip up to this point?

   ____

7. On the final leg of their return trip, they traveled a total distance of 750 miles. After filling up the tank once, they used an additional $\frac{1}{2}$ tank of gas. How many miles did they average per gallon on $1\frac{1}{2}$ tanks of gas?

   ____

Exercise 91 Application Planning a Vacation

Name ______________________________ Date ______________

Bill and Stacy are planning their yearly vacation. Answer the following questions.

1. Stacy called the airlines and found that a one-way airline ticket from Los Angeles to New York City was $237 based on a round-trip purchase. The flight takes about five hours each way. They can travel by train for about two-thirds the cost of the airfare. How much would the round-trip train tickets from Los Angeles to New York City cost for both Bill and Stacy?

   ____

2. After a week in New York, they could fly to Miami and take a cruise. The one-way fare to Miami is $138 per person from New York. The cruise would cost $435 per person. What would the round trip airfare from New York to Miami and the Caribbean cruise cost for both Bill and Stacy?

   ____

3. Bill and Stacy may tour New England to see the historic sights. The tour lasts two weeks and costs $135 per person, not including meals or accommodations. The hotels cost $23 per person per night, and meals cost on the average of $16 per day per person. How much should Bill and Stacy expect to spend altogether?

   ____

4. Bill and Stacy may take a tour to Niagara Falls. From New York City they can take a train at a cost of $67 each way per person. A bus costs $98 each round-trip. How much would Bill and Stacy save if they were to take the bus instead of the train?

   ____

5. The accommodations in Niagara Falls range from $47 per person per night to $168 per couple per night. What is the difference per couple in the hotel rates in Niagara Falls?

   ____

Exercise 92 Synthesis — Estimating Trip Costs

Name ______________________________ Date ______________

Kelly and Karen are planning a three-week vacation along the West Coast of the United States and into Canada. They are planning to drive at least 400 miles each day. Answer the following questions.

1. To budget money for gas, Kelly has decided to keep track of the gas she uses during an average week. Her gas tank holds 18 gallons of gas. She can travel 441 miles on a full tank. How many miles does she average per gallon? If the cost of one gallon of gas is $1.48, find the cost to fill Kelly's gas tank from empty.

   ____

2. Karen wants to find inexpensive lodging while on their trip. West Coast Motels has a chain along the coast from San Diego to Seattle. Kelly and Karen can purchase a one-week package for $161 per person. How much would the lodging cost them altogether for the entire vacation if they stayed at West Coast Motels? How much is that per day, per person?

   ____

3. Karen and Kelly want to apply for passports. The cost of a passport is $67 each. How much will they pay altogether for their passports?

   ____

4. Meals are the most difficult part of a trip to plan. They are going to estimate the cost of meals for the entire trip. They have figured meals for one day will cost them each $20. Circle one of the following statements that would best solve this problem.

   a. $20 \times 7 \times 3$ b. $20 \times 2 \times 21$ c. $20 \times 2 \times 7$ d. $20 \times 20 \times 21$

5. In Seattle, there is a boat tour around Puget Sound. The cost of the two-hour boat tour is $38.50 per person. If the current sales tax rate in Washington is 6.5%, how much will both tickets cost including tax for this tour? How much is that per hour?

   ____